STRUCTURES AND SOLID BODY MECHANICS

GENERAL EDITOR: B. G. NEAL

Introduction to the Theory of Shells

Introduction to the Theory of Shells

CLIVE L DYM

Department of Civil Engineering

Carnegie-Mellon University, Pittsburgh, Pennsylvania

PERGAMON PRESS

OXFORD · NEW YORK

TORONTO · SYDNEY · BRAUNSCHWEIG

Pergamon Press Ltd., Headington Hill Hall, Oxford

Pergamon Press Inc., Maxwell House, Fairview Park, Elmsford, New York 10523

Pergamon of Canada Ltd., 207 Queen's Quay West, Toronto 1

Pergamon Press (Aust.) Pty. Ltd., 19a Boundary Street,
Rushcutters Bay, N.S.W. 2011, Australia

Vieweg & Sohn GmbH, Burgplatz 1, Braunschweig

First Edition 1974

Library of Congress Cataloging in Publication Data

Dym, Clive L
Introduction to the theory of shells.

(Structures and solid body mechanics)
Bibliography: p.
1. Elastic plates and shells. I. Title
QA935.D89 1974 624'.1776 73-13563
ISBN 0-08-017784-0
ISBN 0-08-017785-9 (pbk.)

Printed in Great Britain by Bell and Bain & Co Ltd, Glasgow

לכבוד
יצחק בן אלקסנדר זישא הלוי,
חנה בת אליעזר ליפא.

Contents

Preface

THIS text is meant to provide a brief introduction to the foundations of shell theory, and to some of the important problems that can be tackled within the framework of shell theory. It is in no sense a complete discussion of the theory or its applications. Rather, the text represents a one-semester beginning for students with a reasonable (first course) background in elasticity theory. Hopefully a student who has absorbed this material will be confident in his ability to read and understand the current technical literature in the area.

In this context I should mention that most of the material presented here (and much, much more that is not) has appeared in other places. I have been strongly influenced by the excellent text of H. Kraus and by the outstanding monograph of V. V. Novozhilov, especially in constructing Chapters II and III. My own contribution, as it were, is in the selective distillation and in the ordering required to produce a useful *short* introduction. A bibliography of the major texts and of some of the papers that I have found illuminating will be found at the end of the text.

I should like to express my gratitude to Dr. Steven J. Fenves for providing me with the courage to embark on this endeavor. And, finally, I wish to acknowledge my great debt to Miss Margaret Hall, who has converted my script to typescript with patience, tact, and a great deal of skill.

Pittsburgh CLIVE L. DYM

CHAPTER I

Preludes

THIS text will be concerned with the analysis of structures which physically have three dimensions—as they must—but which can be modeled as two-dimensional surfaces. This is done for reasons of simplicity, for general (three-dimensional) solutions to elasticity problems are not easy to come by. As a simple example of a shell structure, we will consider the Lamé problem and show how a simple membrane theory may be derived therefrom.

As an indication of the approach, we will also examine a development of beam theory, i.e., we will demonstrate the process of modeling a beam in terms of the deflection of a curve.

What is a shell? To quote Flügge,[1] a shell is the "... materialization of a curved surface". So it is, in definition, strictly a matter of *geometry*, and not of material, e.g., a parachute, a concrete roof, a bubble, or even the surface of a liquid can all be treated as shells.

I-1. THE LAMÉ PROBLEM

Now the Lamé solution, for an infinitely long, axisymmetric hollow solid, for $a \leqq r \leqq b$ with the stress boundary conditions

$$\sigma_r\big|_{r=a} = -P_i, \quad \sigma_r\big|_{r=b} = -P_o$$

yields the following stress distribution (see Timoshenko and Goodier)

$$\sigma_r = \frac{a^2b^2(P_o-P_i)}{b^2-a^2}\frac{1}{r^2}+\frac{P_ia^2-P_ob^2}{b^2-a^2}$$

$$\sigma_\theta = -\frac{a^2b^2(P_o-P_i)}{b^2-a^2}\frac{1}{r^2}+\frac{P_ia^2-P_ob^2}{b^2-a^2}.$$

[1] We shall refer to books and papers listed in the Bibliography simply by the author, unless there is some ambiguity.

Let $R = \dfrac{a+b}{2}$, $h = b-a$ so that $a = R - \dfrac{h}{2}$, $b = R+\dfrac{h}{2}$.

Also, let $r = R+z$ so that $-\dfrac{h}{2} \leqq z \leqq \dfrac{h}{2}$.

Clearly R is the mean radius of the shell, and h is the wall thickness. Now we render dimensionless the length quantities with respect to R, and take the outer pressure P_o as zero. Then,

$$\sigma_r = -\frac{P_i(a/R)^2(b/R)^2}{[(b/R)^2-(a/R)^2](r/R)^2}+\frac{P_i(a/R)^2}{(b/R)^2-(a/R)^2}$$

$$\sigma_\theta = +\frac{P_i(a/R)^2(b/R)^2}{[(b/R)^2-(a/R)^2](r/R)^2}+\frac{P_i(a/R)^2}{(b/R)^2-(a/R)^2}.$$

Thus

$$\sigma_r = \frac{P_i(a/R)^2}{(b/R)^2-(a/R)^2}\left[\frac{(r/R)^2-(b/R)^2}{(r/R)^2}\right]$$

$$\sigma_\theta = \frac{P_i(a/R)^2}{(b/R)^2-(a/R)^2}\left[\frac{(r/R)^2+(b/R)^2}{(r/R)^2}\right].$$

Let $m = \dfrac{h}{2R}$, $x = \dfrac{z}{R}$. For *thin shells* $2m = \dfrac{h}{R} \ll 1$, and $-m \leqq x \leqq m$

so that $x \ll 1$ too. Note that the definition of a thin shell, in terms of the thickness-to-radius ratio, h/R, will be a recurrent theme in our development.

Now

$$(b/R)^2-(a/R)^2 = (1+m)^2-(1-m)^2 = 4m$$

$$(a/R)^2 = (1-m)^2;\ (r/R)^2 = (1+x)^2$$

$$(r/R)^2-(b/R)^2 = (1+x)^2-(1+m)^2 = 2(x-m)+\text{h.o.}$$

$$(r/R)^2+(b/R)^2 = 2+\text{h.o.}$$

Therefore we can write the stresses as

$$\sigma_r \cong \frac{P_i(1-m)^2}{4m}\left[\frac{2(x-m)}{(1+x)^2}\right]$$

$$\sigma_\theta \cong \frac{P_i(1-m)^2}{4m}\left[\frac{2}{(1+x)^2}\right].$$

Thus in the limit, for very thin shells,

$$\sigma_r \cong \frac{P_iR}{2h}[0(h/R)](2) = \frac{P_iR}{h}[0(h/R)]$$

$$\sigma_\theta \cong \frac{P_iR}{2h}[0(1)](2) = \frac{P_iR}{h}[0(1)].$$

Thus $\sigma_r/\sigma_\theta = 0(h/R)$, which is very small for thin shells. Thus, the *normal stress through the thickness is considerably smaller (for $h/R \ll 1$) than the in-plane stress.*

We can compare this to well-known membrane results, i.e.,

$$\sigma_\theta = \frac{P_iR}{h}.$$

This assumes no bending!! Thus, our limiting case as $h \to 0$ matches the membrane solution. This implies that very thin shells are membranes, but that bending effects may be important for thicker shells. But it is not difficult to imagine that very thick shells will not bend much either. Thus we shall have to deal with the question of when the membrane theory of shells is adequate, and when it is not.

I-2. A DERIVATION OF BEAM THEORY

We will now develop a simple one-dimensional beam theory from the theory of elasticity. We do this to illustrate the approach we will take in the more complex reduction of three-dimensional elasticity theory to a one-dimensional beam theory. We begin with the strain energy U_e:

$$U_e = \tfrac{1}{2}\int_V \tau_{ij}\varepsilon_{ij}\,dV.$$

Consider a one-dimensional theory, with the assumptions

(1) $$\tau_{xx} = E\varepsilon_{xx}$$

(2) $$\begin{cases} U_1(x, y, z) = U(x) - z\dfrac{dW(x)}{dx} \\ U_2(x, y, z) = 0 \\ U_3(x, y, z) = W(x) \end{cases}$$

(3) $$\int_0^b dy = b; \quad \int_{-h/2}^{h/2} (1, z, z^2)dz = (h, 0, h^3/12).$$

These equations reflect the assumptions of (1) a one-dimensional stress–strain relation, (2) a simple kinematic model, based on the Euler–Bernoulli hypothesis, that writes the deformation of greatest interest in terms of the deflection $W(x)$ of the neutral surface, $z = 0$, and (3) the neutral surface being symmetrically located with respect to the beam thickness.

Then the strain energy can be written as

$$U_e = \int_0^L [\tfrac{1}{2}EA[U'(x)]^2 + \tfrac{1}{2}EI[W''(x)]^2]dx.$$

If we added a potential of a transverse loading $q(x)$, and applied the principle of minimum potential energy (see, for example, Dym and Shames), we would find that the bending problem is governed by the familiar equation

$$(EIW'')'' = q(x).$$

The stretching problem involving $U(x)$ is found to be completely uncoupled from the bending problem, that is, the stretching and bending problems for beams are two independent, disjoint problems. This is primarily a consequence of assumptions (2). But it is most important to note that we have obtained a simplified theory that *represents the bending of a solid beam in terms of the simple deflection $W(x)$ of a straight line.*

In an analogous fashion we will develop the theory of shells, a three-dimensional body with a "thin" dimension, using the theory of surfaces.

Also note that for beams we can define *stress resultants*

$$N = \int_{-h/2}^{h/2} b\tau_{xx} dz = EAU'(x); \quad \tau_{xx}^{\text{memb}} = N/A$$

$$M = \int_{-h/2}^{h/2} b\tau_{xx} z dz = -EIW''(x); \quad \tau_{xx}^{\text{bend}} = Mz/I.$$

Thus again we see that for linear beam theory we have complete *uncoupling* of the stretching [τ_{xx}^{memb}, N, $U(x)$] and the bending [τ_{xx}^{bend}, M, $W(x)$] phenomena.

In the general theory of shells this uncoupling does not occur, i.e., membrane and bending resultants usually depend on the same in-plane *and* out-of-plane displacements.

However, the idea of a linear variation of the displacements through the thickness is used consistently. This linear variation of displacements, incidentally, will not in general lead to the simple stress resultants given above, because of trapezoidal effects in the measurement of area on a curved shell wall.

CHAPTER II

The Theory of Surfaces

We will develop in this chapter the elements of *the theory of surfaces*, following the exposition of Kraus. For additional references, one may consult Novozhilov's monograph on shell theory, or the works of Stoker and Kreyszig on differential geometry.

II-1. THE FIRST FUNDAMENTAL FORM

Every surface S may be defined, in a rectangular coordinate system x_1, x_2, x_3, in terms of two parameters α_1 and α_2:

$$x_i = f_i(\alpha_1, \alpha_2), \quad i = 1, 2, 3 \tag{1}$$

where the f_i are single-valued and continuous functions of the *curvilinear coordinates of the surface* α_1 *and* α_2. By fixing one of the α_k and varying the other, we get a family of *parametric curves* of the surface. The distance from the origin to point P on the surface (Fig. 1) can be given as

$$\mathbf{r} = \mathbf{r}(\alpha_1, \alpha_2) = f_i(\alpha_1, \alpha_2)\mathbf{e}_i \tag{2}$$

where $\mathbf{e}_i$ are unit vectors parallel to the x_i. If one moves to a neighboring point P' on the surface, then

$$d\mathbf{r} = \frac{\partial \mathbf{r}}{\partial \alpha_1} d\alpha_1 + \frac{\partial \mathbf{r}}{\partial \alpha_2}(d\alpha_2) \equiv \mathbf{r}_{,1} d\alpha_1 + \mathbf{r}_{,2} d\alpha_2 = \mathbf{r}_{,i} d\alpha_i. \tag{3}$$

Thus the magnitude of the distance moved on the surface, squared, is

$$(ds)^2 = d\mathbf{r} \cdot d\mathbf{r} = \mathbf{r}_{,i}\mathbf{r}_{,j} d\alpha_i d\alpha_j \equiv E(d\alpha_1)^2 + 2F(d\alpha_1 d\alpha_2) + G(d\alpha_2)^2 \tag{4}$$

where

$$E = \mathbf{r}_{,1} \cdot \mathbf{r}_{,1}, \quad F = \mathbf{r}_{,1} \cdot \mathbf{r}_{,2}, \quad G = \mathbf{r}_{,2} \cdot \mathbf{r}_{,2}. \tag{5}$$

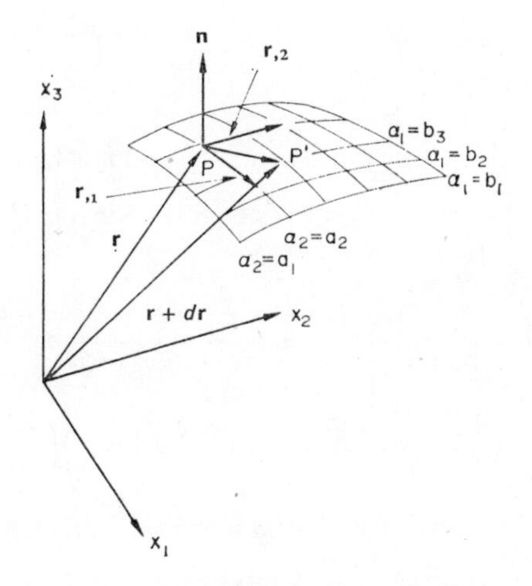

FIG. 1.
Moving from a point P (**r**) to a point P' (**r**$+d$**r**) infinitesmally close to P, on surface.

Then equation (4), with the coefficients defined by equation (5), is called the *first fundamental form* of the surface. The quantities F, E, G are called the *first fundamental magnitudes*. Along the parametric curves themselves, the differential length of arc takes the form

$$\begin{aligned} ds_1 &= \sqrt{(E)}\, d\alpha_1 \qquad \text{(along curve } \alpha_2 = \text{const)} \\ ds_2 &= \sqrt{(G)}\, d\alpha_2 \qquad \text{(along curve } \alpha_1 = \text{const).} \end{aligned} \tag{6}$$

For an *orthogonal net*, since $\mathbf{r}_{,1}$ and $\mathbf{r}_{,2}$ are tangent to curves of constant α_2 and α_1 respectively, it follows that $F = 0$, and that

$$(ds)^2 = A_1^2(d\alpha_1)^2 + A_2^2(d\alpha_2)^2 \tag{7}$$

where $\quad A_1^2 = E \quad \text{and} \quad A_2^2 = G.$

II-2. CURVATURE AND THE SECOND FUNDAMENTAL FORM

At every point P there exists a unit normal vector $\mathbf{n}(\alpha_1, \alpha_2)$ which is normal to $\mathbf{r}_{,1}$, $\mathbf{r}_{,2}$ and to the tangent plane at P that contains the $\mathbf{r}_{,i}$.

Hence

$$\mathbf{n}(\alpha_1, \alpha_2) = \frac{\mathbf{r}_{,1} \times \mathbf{r}_{,2}}{|\mathbf{r}_{,1} \times \mathbf{r}_{,2}|}. \tag{8}$$

Now if θ is the angle between the vectors $\mathbf{r}_{,1}$ and $\mathbf{r}_{,2}$, then

$$|\mathbf{r}_{,1} \times \mathbf{r}_{,2}| = |\mathbf{r}_{,1}|\,|\mathbf{r}_{,2}| \sin\theta, \quad \mathbf{r}_{,1} \cdot \mathbf{r}_{,2} = |\mathbf{r}_{,1}|\,|\mathbf{r}_{,2}| \cos\theta$$

so that

$$\cos\theta = \frac{\mathbf{r}_{,1} \cdot \mathbf{r}_{,2}}{|\mathbf{r}_{,1}|\,|\mathbf{r}_{,2}|} = \frac{\mathbf{r}_{,1} \cdot \mathbf{r}_{,2}}{\sqrt{(\mathbf{r}_{,1} \cdot \mathbf{r}_{,1})}\sqrt{(\mathbf{r}_{,2} \cdot \mathbf{r}_{,2})}} \equiv \frac{F}{\sqrt{(EG)}}$$

$$\sin\theta = \sqrt{(1-\cos^2\theta)} = \sqrt{\left(1-\frac{F^2}{EG}\right)} \equiv \sqrt{\left(\frac{EG-F^2}{EG}\right)}$$

and thus the final expression for the *unit normal vector* **n** is

$$\mathbf{n}(\alpha_1, \alpha_2) = \frac{\mathbf{r}_{,1} \times \mathbf{r}_{,2}}{H}, \quad H \equiv \sqrt{(EG-F^2)}. \tag{9}$$

We adopt the convention, henceforth, that the parametric curves will be arranged so that $\mathbf{n}(\alpha_1, \alpha_2)$ is positive when pointing from the concave side of the surface to the convex side.

Now if **t** is the tangent to a curve on the surface, and is the parameter used to define the curve, the *curvature vector* **k** can be defined as $\mathbf{k} = d\mathbf{t}/ds$, which can further be resolved into normal and tangential components, with respect to the surface,

$$\mathbf{k} = \frac{d\mathbf{t}}{ds} = \mathbf{k}_n + \mathbf{k}_t\,. \tag{10}$$

We are interested only in the normal curvature vector $\mathbf{k}_n$. Since $\mathbf{k}_n$ is in the direction of the normal, and since **k** is usually positive from convex to concave, let

$$\mathbf{k}_n = -K_n \mathbf{n}. \tag{11}$$

Here K_n is the *normal curvature*. To find an explicit expression for K_n, note first that $\mathbf{t} \cdot \mathbf{n} = 0$ always, so that $d(\mathbf{t} \cdot \mathbf{n})/ds = 0$, or

$$\frac{d\mathbf{n}}{ds} \cdot \mathbf{t} = -\frac{d\mathbf{t}}{ds} \cdot \mathbf{n}\,. \tag{12}$$

From equation (11) note that

$$(\mathbf{k}_n \cdot \mathbf{n}) = -K_n(\mathbf{n} \cdot \mathbf{n}) = -K_n$$

and from equation (10)

$$\mathbf{k} \cdot \mathbf{n} = \frac{d\mathbf{t}}{ds} \cdot \mathbf{n} = \mathbf{k}_n \cdot \mathbf{n} + \mathbf{k}_t \cdot \mathbf{n} = \mathbf{k}_n \cdot \mathbf{n}$$

so that

$$K_n = -(\mathbf{k}_n \cdot \mathbf{n}) = -\left(\frac{d\mathbf{t}}{ds} \cdot \mathbf{n}\right) = \frac{d\mathbf{n}}{ds} \cdot \mathbf{t}$$

$$\equiv \frac{d\mathbf{n}}{ds} \cdot \frac{d\mathbf{r}}{ds} = \frac{d\mathbf{n} \cdot d\mathbf{r}}{(ds)^2} = \frac{d\mathbf{n} \cdot d\mathbf{r}}{d\mathbf{r} \cdot d\mathbf{r}} \tag{13}$$

where we have used the fact that the tangent to a curve may be found as $\mathbf{t} = d\mathbf{r}/ds$. Then, since

$$d\mathbf{n} = \mathbf{n}_{,i} d\alpha_i, \quad d\mathbf{r} = \mathbf{r}_{,i} d\alpha_i$$

it follows that

$$K_n = \frac{L(d\alpha_1)^2 + 2M(d\alpha_1 d\alpha_2) + N(d\alpha_2)^2}{E(d\alpha_1)^2 + 2F(d\alpha_1 d\alpha_2) + G(d\alpha_2)^2} \tag{14}$$

where the numerator of K_n is the *second fundamental form*, containing the *second fundamental magnitudes*:

$$L = \mathbf{n}_{,1} \cdot \mathbf{r}_{,1}, \quad 2M = \mathbf{n}_{,1} \cdot \mathbf{r}_{,2} + \mathbf{n}_{,2} \cdot \mathbf{r}_{,1}, \quad N = \mathbf{n}_{,2} \cdot \mathbf{r}_{,2}. \tag{15}$$

Since it is further true that $\mathbf{n} \cdot \mathbf{r}_{,1} = 0$, by differentiation one can also show that

$$L = -\mathbf{n} \cdot \mathbf{r}_{,11}, \quad M = -\mathbf{n} \cdot \mathbf{r}_{,12}, \quad N = -\mathbf{n} \cdot \mathbf{r}_{,22} \tag{16}$$

where $\mathbf{r}_{,ij} = \partial^2 \mathbf{r}/\partial\alpha_i \partial\alpha_j$.

We now seek directions $(d\alpha_2/d\alpha_1) \equiv \lambda$ for which the curvature is a maximum or a minimum, i.e., for

$$K_n(\lambda) = \frac{L + 2M\lambda + N\lambda^2}{E + 2F\lambda + G\lambda^2}$$

we seek λ such that $dK_n/d\lambda = 0$. After some manipulation we find that the extreme value of $K_n(\lambda)$ is given by

$$K_n = \frac{M+N\lambda}{F+G\lambda} = \frac{L+M\lambda}{E+F\lambda} \tag{17}$$

where λ is a root of

$$(MG-NF)\lambda^2+(LG-NE)\lambda+(LF-ME) = 0. \tag{18}$$

There are two roots to this quadratic, λ_1 and λ_2, corresponding to the maximum curvature and the minimum curvature. They are denoted as *principal curvatures* K_1 *and* K_2, with *principal radii of curvature* R_1 *and* R_2, where $R_i = K_i^{-1}$.

We can show that the principal curvature directions are orthogonal to each other. Let θ be the angle between the directions $d\alpha_2/d\alpha_1$ and $\delta\alpha_2/\delta\alpha_1$. Then along these directions a differential change in position is given by

$$d\mathbf{r} = \mathbf{r}_{,i}d\alpha_i, \quad \delta\mathbf{r} = \mathbf{r}_{,i}\delta\alpha_i$$

so that from the definition of a scalar product

$$\cos\theta = \frac{d\mathbf{r}\cdot\delta\mathbf{r}}{|d\mathbf{r}|\,|\delta\mathbf{r}|} \equiv E\frac{d\alpha_1}{ds}\frac{\delta\alpha_1}{\delta s}+F\left(\frac{d\alpha_1}{ds}\frac{\delta\alpha_2}{\delta s}+\frac{d\alpha_2}{ds}\frac{\delta\alpha_1}{\delta s}\right)+G\frac{d\alpha_2}{ds}\frac{\delta\alpha_2}{\delta s}.$$

For orthogonality, $\theta = \pi/2$, and identifying λ_1 with $d\alpha_2/d\alpha_1$ and λ_2 with $\delta\alpha_2/\delta\alpha_1$, we must have

$$E+F(\lambda_1+\lambda_2)+G\lambda_1\lambda_2 = 0.$$

Now from the quadratic [equation (18)] for the roots λ one can find that

$$\lambda_1+\lambda_2 = -\frac{(LG-NE)}{(MG-NF)}$$

$$\lambda_1\lambda_2 = \frac{(LF-ME)}{(MG-NF)}.$$

Upon substitution we find that the orthogonality condition is satisfied!!

Now if the lines of principal curvature are taken as the parametric lines (curves) of the surface, then equation (18) must be satisfied for both $d\alpha_2/d\alpha_1 = 0$ and $d\alpha_1/d\alpha_2 = 0$, which implies that

$$LF - ME = 0 \quad \text{and} \quad MG - NF = 0.$$

Since the lines of curvature are orthogonal, $F = 0$. Since it can be shown[2] that $EG - F^2 > 0$, it follows that neither E or G can vanish if $F = 0$. Thus it must be true that $M = 0$. Thus the parametric lines are lines of principal curvature if

$$F = M = 0. \tag{19}$$

Then it follows from equation (14), with $d\alpha_1 = 0$ and $d\alpha_2 = 0$ in turn, that the curvatures are

$$K_1 = \frac{1}{R_1} = \frac{L}{E}, \quad K_2 = \frac{1}{R_2} = \frac{N}{G}. \tag{20}$$

In the subsequent work we will assume that equation (19) is satisfied, and that the lines of principal curvature of the reference surface are the parametric lines.

II-3. THE GAUSS–CODAZZI CONDITIONS AND THE FUNDAMENTAL THEOREM

For these principal lines of curvature, in developing the fundamental theorem of the theory of surfaces, we will need certain vector identities. First we will define some unit vectors tangent to the α_1 and α_2 lines and a unit vector that is normal to the surface:

$$\mathbf{t}_1 = \frac{\mathbf{r}_{,1}}{|\mathbf{r}_{,1}|} = \frac{\mathbf{r}_{,1}}{A_1}$$

$$\mathbf{t}_2 = \frac{\mathbf{r}_{,2}}{|\mathbf{r}_{,2}|} = \frac{\mathbf{r}_{,2}}{A_2} \tag{21}$$

$$\mathbf{n} = \mathbf{t}_1 \times \mathbf{t}_2 = \frac{\mathbf{r}_{,1} \times \mathbf{r}_{,2}}{A_1 A_2}.$$

[2] $EG - F^2 = (\mathbf{r}_{,1} \cdot \mathbf{r}_1)(\mathbf{r}_{,2} \cdot \mathbf{r}_{,2}) - (\mathbf{r}_{,1} \cdot \mathbf{r}_{,2})^2$.
If $|\mathbf{r}_{,i}| \equiv r_i$ then $\mathbf{r}_{,i} \cdot \mathbf{r}_{,i} = r_i^2$.
And $\mathbf{r}_{,1} \cdot \mathbf{r}_{,2} = r_1 r_2 \cos\phi$ where $\phi = (\mathbf{r}_{,1}, \mathbf{r}_{,2})$.
Then

$$EG - F^2 = r_1^2 r_2^2 - (r_1 r_2 \cos\phi)^2 = r_1^2 r_2^2 - r_1^2 r_2^2 \cos^2\phi$$
$$= (r_1 r_2)^2 \sin^2\phi > 0 \quad \text{if} \quad \phi \neq 0.$$

Now since the $\mathbf{n}_{,i} = \partial \mathbf{n}/\partial \alpha_i$ are perpendicular to $\mathbf{n}$, they must lie in the plane of $\mathbf{t}_1$ and $\mathbf{t}_2$. It must thus be true, for example, that

$$\mathbf{n}_{,1} = a\mathbf{t}_1 + b\mathbf{t}_2$$

so that

$$\mathbf{t}_1 \cdot \mathbf{n}_{,1} = \frac{\mathbf{r}_{,1} \cdot \mathbf{n}_{,1}}{A_1} = a\mathbf{t}_1 \cdot \mathbf{t}_1 + b\mathbf{t}_2 \cdot \mathbf{t}_1 = a$$

$$\mathbf{t}_2 \cdot \mathbf{n}_{,1} = \frac{\mathbf{r}_{,2} \cdot \mathbf{n}_{,1}}{A_2} = a\mathbf{t}_1 \cdot \mathbf{t}_2 + b\mathbf{t}_2 \cdot \mathbf{t}_2 = b$$

or

$$a = \frac{\mathbf{r}_{,1} \cdot \mathbf{n}_{,1}}{A_1} = \frac{L}{A_1}$$
$$b = \frac{\mathbf{r}_{,2} \cdot \mathbf{n}_{,1}}{A_2} = \frac{M}{A_2} = 0. \tag{22}$$

Hence

$$\mathbf{n}_{,1} = \frac{L}{A_1}\mathbf{t}_1 = \frac{E}{R_1 A_1}\mathbf{t}_1 = \frac{A_1}{R_1}\mathbf{t}_1 . \tag{23a}$$

In a similar way,

$$\mathbf{n}_{,2} = \frac{A_2}{R_2}\mathbf{t}_2 \tag{23b}$$

Now we proceed to find derivatives of $\mathbf{t}_1$ and $\mathbf{t}_2$. First, for continuous $\mathbf{r}$, it must be true that $\mathbf{r}_{,12} = \mathbf{r}_{,21}$, so that

$$(A_1\mathbf{t}_1)_{,2} = (A_2\mathbf{t}_2)_{,1} \tag{24}$$

or

$$\mathbf{t}_{2,1} = \frac{1}{A_2}[A_1\mathbf{t}_{1,2} + \mathbf{t}_1 A_{1,2} - \mathbf{t}_2 A_{2,1}] . \tag{24a}$$

We then use the fact that $\mathbf{t}_{1,1}$ must lie in the plane formed by $\mathbf{t}_2$ and $\mathbf{n}$. Thus we can say that

$$\mathbf{t}_{1,1} = c\mathbf{n} + d\mathbf{t}_2$$

so that it follows that

$$\mathbf{n}\cdot\mathbf{t}_{1,1} = c, \quad \mathbf{t}_2\cdot\mathbf{t}_{1,1} = d.$$

Now since $\mathbf{n}\cdot\mathbf{t}_1 = 0$, differentiation yields

$$c = -\mathbf{n}_{,1}\cdot\mathbf{t}_1 = -\frac{A_1}{R_1}. \tag{25}$$

From equation (24a) and the above, since $\mathbf{t}_2\cdot\mathbf{t}_1 = 0$,

$$d = -\mathbf{t}_1\cdot\mathbf{t}_{2,1} = -\frac{\mathbf{t}_1}{A_2}\cdot[A_1\mathbf{t}_{1,2}+\mathbf{t}_1 A_{1,2}-\mathbf{t}_2 A_{2,1}]$$

$$= -\frac{A_{1,2}}{A_2}. \tag{26}$$

Thus, finally,

$$\mathbf{t}_{1,1} = -\frac{A_1}{R_1}\mathbf{n}-\frac{1}{A_2}\frac{\partial A_1}{\partial \alpha_2}\mathbf{t}_2. \tag{27a}$$

In a similar way

$$\mathbf{t}_{1,2} = \frac{1}{A_1}\frac{\partial A_2}{\partial \alpha_1}\mathbf{t}_2 \tag{27b}$$

$$\mathbf{t}_{2,1} = \frac{1}{A_2}\frac{\partial A_1}{\partial \alpha_2}\mathbf{t}_1 \tag{27c}$$

$$\mathbf{t}_{2,2} = -\frac{A_2}{R_2}\mathbf{n}-\frac{1}{A_1}\frac{\partial A_2}{\partial \alpha_1}\mathbf{t}_1. \tag{27d}$$

Now we shall derive the three differential equations—*the Gauss–Codazzi* conditions—that relate the four hitherto arbitrary parameters A_1, A_2, R_1 and R_2. The conditions will be found from the requirement that the mixed second derivatives of the unit vectors are equal, implying continuity of the second derivatives of the unit vectors. Starting with, for example,

$$\mathbf{n}_{,12} = \mathbf{n}_{,21}$$

we need

$$\left(\frac{A_1}{R_1}\mathbf{t}_1\right)_{,2} = \left(\frac{A_2}{R_2}\mathbf{t}_2\right)_{,1}$$

which, in view of equations (27), yields

$$\mathbf{t}_1\left[-\frac{A_{1,2}}{R_2}+\left(\frac{A_1}{R_1}\right)_{,2}\right]+\mathbf{t}_2\left[\frac{A_{2,1}}{R_1}-\left(\frac{A_2}{R_2}\right)_{,1}\right]=0\,.$$

This vector equation will be true only if the brackets vanish. Thus

$$\frac{1}{R_2}\,A_{1,2}=\left(\frac{A_1}{R_2}\right)_{,2};\quad \frac{1}{R_1}\,A_{2,1}=\left(\frac{A_2}{R_2}\right)_{,1}\,. \tag{28}$$

Equations (28) are the *Codazzi conditions.*

Proceeding from $\mathbf{t}_{1,12}=\mathbf{t}_{1,21}$, we find only one new condition, the *Gauss condition*

$$\left(\frac{1}{A_1}\,A_{2,1}\right)_{,1}+\left(\frac{1}{A_2}\,A_{1,2}\right)_{,2}=-\frac{A_1A_2}{R_1R_2}\,. \tag{29}$$

Since four quantities can be related by no more than three homogeneous equations, if they are to possess nontrivial solutions, the result of examining $\mathbf{t}_{2,12}=\mathbf{t}_{2,21}$ yields—as expected—nothing new.

We can now state, for lines of principal curvature used as parametric lines, the *fundamental theorem of the theory of surfaces*:

If E, G, L and N **are** *given as functions of the real curvilinear coordinates* α_1 *and* α_2, *and are sufficiently differentiable and satisfy the Gauss–Codazzi conditions while* $E>0$ *and* $G>0$, *then there exists a real surface which has as its first and second fundamental forms*

$$I=R(d\alpha_1)^2+G(d\alpha_2)^2$$

$$II=L(d\alpha_1)^2+N(d\alpha_2)^2\,.$$

This surface is uniquely determined except for its position in space.

Remember that the above development is valid only for lines of principal curvature. We also note that the Gauss–Codazzi conditions amount to compatibility equations for surfaces.

II-4. THE SURFACE OF REVOLUTION

To illustrate the application of some of these results, let us consider the geometry of a surface of revolution, shown in Fig. 2.

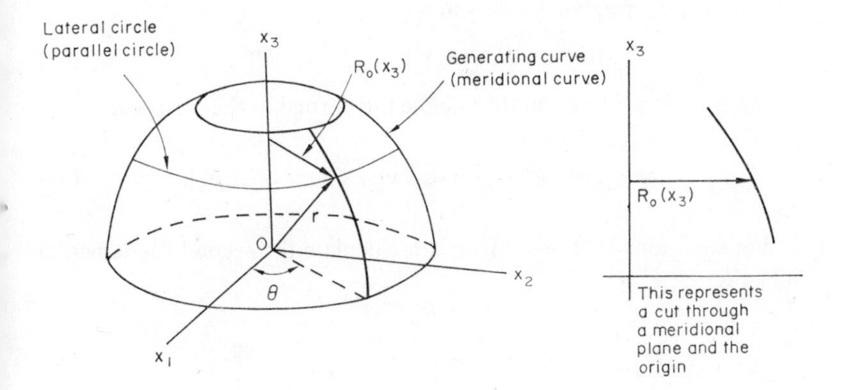

FIG. 2.

We generate a surface of revolution by rotating a plane curve (*the meridional curve*) about an axis of revolution. It is clear that the position vector can be written as

$$\mathbf{r}(x_3, \theta) = R_0(x_3) \cos\theta \, \mathbf{e}_1 + R_0(x_3) \sin\theta \, \mathbf{e}_2 + x_3 \mathbf{e}_3 . \tag{30}$$

To calculate the fundamental forms and magnitudes, let us associate α_1 with x_3 and α_2 with θ. Then

$$\begin{aligned} \mathbf{r}_{,1} &= R_0' \cos\theta \, \mathbf{e}_1 + R_0' \sin\theta \, \mathbf{e}_2 + \mathbf{e}_3 \\ \mathbf{r}_{,2} &= -R_0 \sin\theta \, \mathbf{e}_1 + R_0 \cos\theta \, \mathbf{e}_2 . \end{aligned} \tag{31}$$

Then using equations (5) we can calculate the first fundamental magnitudes

$$\begin{aligned} E &= 1 + (R_0')^2 \\ F &= 0 \\ G &= R_0^2 \\ H &= \sqrt{(EG - F^2)} = R_0 \sqrt{[1 + (R_0')^2]} . \end{aligned} \tag{32}$$

Thus, with $F = 0$, an orthogonal net is formed by α_1, α_2 (or x_3, θ),

$$A_1 = \sqrt{E} = \sqrt{[1 + (R_0')^2]}, \quad A_2 = \sqrt{G} = R_0 \tag{33}$$

so that the first fundamental form is

$$(ds)^2 = [1+(R_0')^2](dx_3)^2 + R_0^2(d\theta)^2 . \tag{34}$$

Using equation (9) we can then define the normal to the surface as

$$\mathbf{n}(x_3, \theta) = -\frac{R_0}{H}[\cos\theta\,\mathbf{e}_1 + \sin\theta\,\mathbf{e}_2 - R_0'\mathbf{e}_3]. \tag{35}$$

Using equations (15) or (16) we can calculate the second fundamental magnitudes,

$$\begin{aligned} L &= R_0 R_0''/H \\ M &= 0 \\ N &= R_0^2/H . \end{aligned} \tag{36}$$

Thus, since both $F = M = 0$, the meridians and parallels are lines of principal curvature. The principal radii can be calculated from equations (20):

$$\begin{aligned} R_1 &= -[1+(R_0')^2]^{3/2} R_0'' = \frac{1}{K_1} \\ R_2 &= R_0\sqrt{[1+(R_0')^2]} = \frac{1}{K_2}. \end{aligned} \tag{37}$$

It is a straightforward matter to verify that the A_i and R_i defined by equations (33) and (37) satisfy the Gauss–Codazzi conditions.

An alternate set of descriptions for the shell of revolution makes use of a "meridional angle" ϕ, and θ, where the angle ϕ is measured between the axis of revolution and the normal to the surface. If $\alpha_1 \to \phi$, $\alpha_2 \to \theta$, the first fundamental form is

$$(ds)^2 = R_1^2(d\phi)^2 + R_0^2(d\theta)^2 \tag{38}$$

so that $A_1 = R_1$ and $A_2 = R_0$. Here the Gauss condition will require that

$$\frac{dR_0}{d\phi} = R_1 \cos\phi \tag{39}$$

which can easily be verified (see Fig. 3).

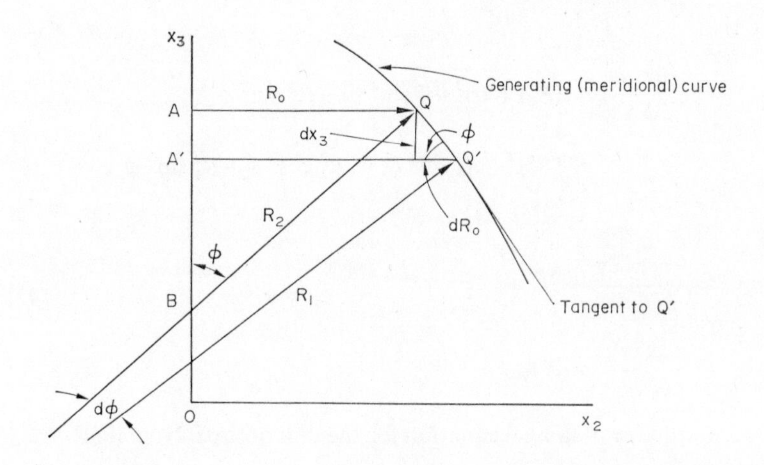

FIG. 3.

Clearly $BQ = R_2$ and $AQ = R_0 = R_2 \sin \phi$ so that

$$dR_0 = A'Q' - AQ = QQ' \cos \phi = R_1 d\phi \cos \phi$$

and so

$$\frac{dR_0}{d\phi} = R_1 \cos \phi \, .$$

We can also derive these forms from the position vector

$$\mathbf{r}(\phi, \theta) = R_0 \cos \theta \, \mathbf{e}_1 + R_0 \sin \theta \, \mathbf{e}_2 + x_3(\phi)\mathbf{e}_3 \tag{40}$$

where $R_0 = R_0(\phi)$, $x_3 = x_3(\phi)$ and

$$dx_3(\phi) = dx_3 = -R_1 \sin \phi \, d\phi \, . \tag{41}$$

Then, verifying equation (41) in Fig. 3, one can calculate

$$\frac{\partial \mathbf{r}}{\partial \phi} = R_0' \cos \theta \, \mathbf{e}_1 + R_0' \sin \theta \, \mathbf{e}_2 + \frac{dx_3}{d\phi} \mathbf{e}_3$$

$$= R_0' \cos \theta \, \mathbf{e}_1 + R_0' \sin \theta \, \mathbf{e}_2 - R_1 \sin \phi \, \mathbf{e}_3$$

$$\frac{\partial \mathbf{r}}{\partial \theta} = -R_0 \sin \theta \mathbf{e}_1 + R_0 \cos \theta \mathbf{e}_2$$

so that

$$\left.\begin{aligned}\frac{\partial \mathbf{r}}{\partial \phi}\cdot\frac{\partial \mathbf{r}}{\partial \phi} &= (R_0')^2(\cos^2\theta+\sin^2\theta)+R_1^2\sin^2\phi \\ &= (R_0')^2+R_1^2\sin^2\phi = R_1^2\cos^2\phi+R_1^2\sin^2\phi \\ &= R_1^2 = A_1^2 = E \\ \frac{\partial \mathbf{r}}{\partial \theta}\cdot\frac{\partial \mathbf{r}}{\partial \theta} &= R_0^2 = A_2^2 = G \\ \frac{\partial \mathbf{r}}{\partial \phi}\cdot\frac{\partial \mathbf{r}}{\partial \theta} &= 0 = F\end{aligned}\right\}. \tag{42}$$

We will now obtain an equation for the unit normal. Note that

$$H = \sqrt{(EG-F^2)} = A_1A_2 = R_0R_1$$

and that

$$\frac{\partial \mathbf{r}}{\partial \phi}\times\frac{\partial \mathbf{r}}{\partial \theta} = \begin{vmatrix}\mathbf{e}_1 & \mathbf{e}_2 & \mathbf{e}_3 \\ R_0'\cos\theta & R_0'\sin\theta & -R_1\sin\phi \\ -R_0\sin\theta & R_0\cos\theta & 0\end{vmatrix}$$

$$= +R_0R_1(\sin\phi\cos\theta\mathbf{e}_1+\sin\phi\sin\theta\mathbf{e}_2)+R_0R_0'\mathbf{e}_3$$

so that

$$\mathbf{n}(\phi,\theta) = +\sin\phi\cos\theta\mathbf{e}_1+\sin\theta\sin\theta\mathbf{e}_2+R_0'/R_1\mathbf{e}_3. \tag{43}$$

We shall also need the second derivatives of **r**,

$$\frac{\partial^2 \mathbf{r}}{\partial \phi^2} = R_0''\cos\theta\mathbf{e}_1+R_0''\sin\theta\mathbf{e}_2-(R_1\sin\phi)'\mathbf{e}_3$$

$$\frac{\partial^2 \mathbf{r}}{\partial \phi\partial \theta} = -R_0'\sin\theta\mathbf{e}_1+R_0'\cos\theta\mathbf{e}_2$$

$$\frac{\partial^2 \mathbf{r}}{\partial \theta^2} = -R_0\cos\theta\mathbf{e}_2-R_0\sin\theta\mathbf{e}_2\,.$$

Then

$$L = -\mathbf{n}\cdot\frac{\partial^2\mathbf{r}}{\partial\phi^2} = -R_0''\sin\phi+(R_0'/R_1)(R_1\sin\phi)' = +R_1$$

$$M = -\mathbf{n}\cdot\frac{\partial^2\mathbf{r}}{\partial\phi\partial\theta} = 0 \tag{44}$$

$$N = -\mathbf{n}\cdot\frac{\partial^2\mathbf{r}}{\partial\theta^2} = +R_0\sin\phi\,.$$

From equations (20) the principal curvatures are seen to be

$$K_1 = \frac{L}{E} = +\frac{1}{R_1}, \quad K_2 = \frac{N}{G} = +\frac{\sin\phi}{R_0} = +\frac{1}{R_2} \tag{45}$$

so that the principal radii of curvature are

$$R_1^{\text{princip}} = +R_1, \quad R_2^{\text{princip}} = +R_2\,. \tag{46}$$

Thus the R_1, R_2 used from equations (38) to (46) and in Fig. 3 are the principal radii of curvature.

To consider a sphere, for example, in Fig. 3 we could simply take $R_1 = R_2 = a$, $R_0 = a\sin\phi$, $x_3 = a\cos\phi$, and

$$\mathbf{r}(\phi, \theta) = a(\sin\phi\cos\theta\mathbf{e}_1+\sin\phi\sin\theta\mathbf{e}_2+\cos\phi\mathbf{e}_3). \tag{47}$$

This implies that $R_1^{\text{princip}} = R_2^{\text{princip}} = +a.$

II-5. SOME TERMINOLOGY FOR SURFACES

Before closing this chapter, a few more remarks about curvature are in order. If the lines of principal curvature do not coincide with the coordinate (parametric) lines, to actually solve for the principal curvatures we would use equations (17) in the form

$$(EK_n-L)+(FK_n-M)\lambda = 0$$

$$(FK_n-M)+(GK_n-N)\lambda = 0.$$

These two equations are satisfied for a value of λ if the determinant of the coefficients vanishes, i.e.,

$$(EK_n-L)(GK_n-N)-(FK_n-M)^2 = 0$$

which yields a quadratic equation for the principal curvatures,

$$H^2K_n^2-(LG+EN-2FM)K_n+NL-M^2 = 0. \tag{48}$$

As with the quadratic analogy previously used, we see that the sum and products of the principal curvatures are given as

$$K_1+K_2 = \frac{LG+EN-2FM}{H^2} \equiv 2G \tag{49a}$$

$$K_1K_2 = \frac{NL-M^2}{H^2} \equiv K \tag{49b}$$

where $G \equiv$ *mean curvature of the surface*, and $K \equiv$ *Gaussian curvature of the surface*. Having G and K we can then solve for K_1 and K_2.

A surface is said to have *positive curvature if* $K > 0$, and it has *negative curvature if* $K < 0$. We also use the terms *elliptic* ($K > 0$), *hyperbolic* ($K < 0$), and *parabolic* ($K = 0$).

Note that, if the lines of curvature do not coincide with the coordinate lines, then K_1 and K_2 are the curvatures of normal sections in the directions of lines of principal curvature.

For the case where the coordinate lines are the lines of curvature, we can write the Gauss condition (equation (29)) in the form

$$A_1A_2K = -\left(\frac{A_{2,1}}{A_1}\right)_{,1} -\left(\frac{A_{1,2}}{A_2}\right)_{,2} . \tag{50}$$

Hence $K = K(A_1, A_2)$, and if a surface is bent without straining, then $(ds)^2$ is not changed, and thus A_1 and A_2 are not changed. Hence, with no straining, K does not change and it is thus a *bending invariant*.

Also, for a plane, $K = 0$. Thus, $K = 0$ for any surface that can be obtained from a plane without strain. Such surfaces are called *developable surfaces*, and include cylinders and cones.

CHAPTER III

The Construction of a Shell Theory

WE begin now our development of shell theory, starting with the basic postulates, or assumptions.

Recall that in elasticity theory there are three basic sets of equations—equilibrium, kinematic (strain–displacement), and constitutive (Hooke's law). As the solution of three-dimensional elasticity problems is tough, shell theory may be viewed as a two-dimensional sub-set of elasticity valid for certain classes of structures. For this sub-set we shall develop appropriate kinematic, constitutive and equilibrium relations. Our discussion will be limited to static isothermal loading of isotropic shells. Thus it is in part a sub-set of the development of Kraus, for example, who includes the inertia terms, orthotropic elasticity, and thermal stress terms.

III-1. THE BASIC ASSUMPTIONS

There are four principal assumptions that are used to establish thin shell theory. The most basic is that one dimension is considerably smaller than the other two, so we can speak of a *thin* shell. The second is that the shell deflections are assumed to be small. Thirdly, we take the stress in the direction normal to the thin dimension to be negligible. Finally, we assume that a line originally normal to the shell reference surface will remain normal to the deformed reference surface, and unstrained (unextended).

It seems clear that the first postulate is basic to the entire theory, and it makes the other postulates possible and reasonable. As a measure of "thin-ness" we say that the ratio of shell thickness (h) to one of the radii of curvature (R_i) shall be small compared to unity, i.e., $h/R_i \ll 1$. For engineering accuracy the upper limit of this ratio is variously put at 1/10 or 1/20.

The small deflection assumption allows us to refer all equilibrium and kinematic questions to the original, undeformed, reference state of the shell. Thus we need not distinguish, say, between Eulerian and Lagrangian descriptions (see Novozhilov on nonlinear elasticity theory, for example).

The third postulate simply states that σ_n, say, is much smaller than σ_1, σ_2, the normal in-plane stresses, i.e., $\sigma_n \ll \sigma_1, \sigma_2$. (Recall here the limiting results of the Lamé problem given in Chapter I.) Then, for an isotropic, homogeneous elastic solid, the in-plane constitutive law is given by the plane stress equations:

$$\begin{aligned} \varepsilon_1 &= \frac{1}{E}(\sigma_1 - \nu\sigma_2) \\ \varepsilon_2 &= \frac{1}{E}(\sigma_2 - \nu\sigma_1) \end{aligned} \tag{51}$$

or

$$\begin{aligned} \sigma_1 &= \frac{E}{1-\nu^2}(\varepsilon_1 + \nu\varepsilon_2) \\ \sigma_2 &= \frac{E}{1-\nu^2}(\varepsilon_2 + \nu\varepsilon_1). \end{aligned} \tag{52}$$

The final assumption is an analogy to the Euler hypothesis of plane sections remaining plane in beam theory. It is sometimes called the "hairbrush hypothesis". The assumption of the unextended normal implies that all strain components in the direction of the normal to the reference surface vanish. Thus

$$\varepsilon_n = \gamma_{1n} = \gamma_{2n} = 0. \tag{53}$$

Thus, in conjunction with the isotropic Hooke's law, we can write that

$$\tau_{12} = G\gamma_{12} = 2G\varepsilon_{12}, \quad \tau_{1n} = \tau_{2n} = 0. \tag{54}$$

Our next step is to define shell coordinates.

III-2. SHELL COORDINATES

To define *shell coordinates*, note that the assumption regarding the preservation of the normal implies that the displacements must be linear in the thickness coordinate. Thus the behavior of any point in the shell could be determined from the behavior of a corresponding point on a reference surface. It is convenient for our work, though by no means universal, to choose the *middle surface* as the reference surface. Also, we continue to use an (orthogonal) curvilinear coordinate system which coincides with the orthogonal lines of principle curvature of the surface. Thus to an arbitrary point within the shell, define the position vector

$$\mathbf{R}(\alpha_1, \alpha_2, z) = \mathbf{r}(\alpha_1, \alpha_2) + z\mathbf{n}(\alpha_1, \alpha_2) \tag{55}$$

where $\mathbf{r}$ is the position vector to a point on the reference surface, $\mathbf{n}$ is the unit vector normal to the reference surface, and z *a coordinate measured from the reference surface, through the thickness, along* $\mathbf{n}(\alpha_1, \alpha_2)$. The α_i are bounded by some values that define the shell boundaries.

The magnitude of a differential element of length is then given by

$$\begin{aligned}(ds)^2 &= d\mathbf{R}\cdot d\mathbf{R} = (d\mathbf{r}+zd\mathbf{n}+dz\mathbf{n})\cdot(d\mathbf{r}+zd\mathbf{n}+dz\mathbf{n}) \\ &= A_1^2(1+z/R_1)^2(d\alpha_1)^2 + A_2^2(1+z/R_2)^2(d\alpha_2)^2 + (dz)^2. \end{aligned} \tag{56}$$

The verification of equation (56) is shown in Appendix IIIA.

Now the first two terms on the right-hand side of equation (56) can be thought of as the first fundamental form of a surface at a distance z from the reference surface. For such a surface,

$$\begin{aligned}\frac{\partial}{\partial\alpha_2}[A_1(1+z/R_1)] &= \frac{\partial A_1}{\partial\alpha_2} + z\frac{\partial}{\partial\alpha_2}\left(\frac{A_1}{R_1}\right) = \frac{\partial A_1}{\partial\alpha_2} + \frac{z}{R_2}\frac{\partial A_1}{\partial\alpha_2} \\ &= (1+z/R_2)\frac{\partial A_1}{\partial\alpha_2} \\ \frac{\partial}{\partial\alpha_1}[A_2(1+z/R_2)] &= (1+z/R_1)\frac{\partial A_2}{\partial\alpha_1}.\end{aligned} \tag{57}$$

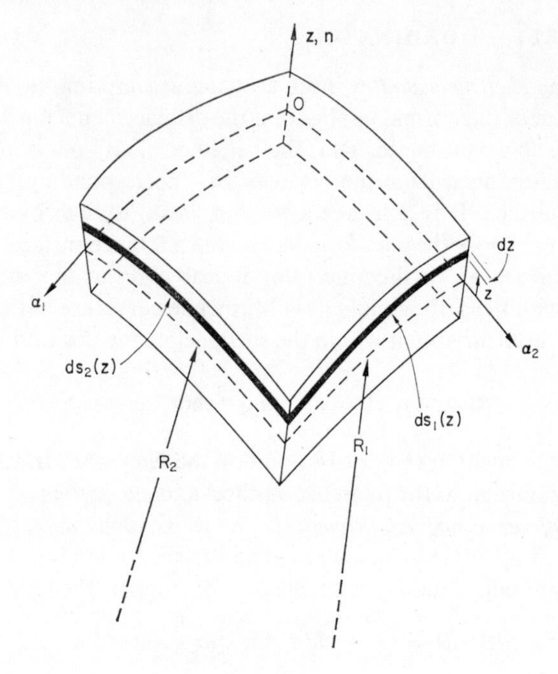

FIG. 4.

In Fig. 4 is displayed a shell (differential) element.

From Fig. 4, and the definition of $(ds)^2$, it is clear that if we isolate a shell element of thickness dz at an altitude z from the middle surface, the lengths of the edges of the element are given by

$$
\begin{aligned}
ds_1(z) &= A_1(1+z/R_1)(d\alpha_1) \\
ds_2(z) &= A_2(1+z/R_2)(d\alpha_2)
\end{aligned}
\tag{58}
$$

and the corresponding area elements of the faces are

$$
\begin{aligned}
d\Sigma_1(z) &= A_1(1+z/R_1)(d\alpha_1)(dz) \\
d\Sigma_2(z) &= A_2(1+z/R_2)(d\alpha_2)(dz).
\end{aligned}
\tag{59}
$$

III-3. STRAIN–DISPLACEMENT RELATIONS

We shall now present the *strain–displacement* relations of shell theory, accepting for now some results from elasticity theory in general curvilinear coordinates.[3] Define first a displacement vector in the shell coordinates,

$$\mathbf{u}(\alpha_1, \alpha_2, z) = U_1(\alpha_1, \alpha_2, z)\mathbf{t}_1 + U_2(\alpha_1, \alpha_2, z)\mathbf{t}_2 + W(\alpha_1, \alpha_2, z)\mathbf{n}\,. \quad (60)$$

Now in an orthogonal curvilinear system, the normal and shearing components of strain are

$$\varepsilon_i = \frac{\partial}{\partial \alpha_i}\left(\frac{U_i}{\sqrt{(g_i)}}\right) + \frac{1}{2g_i}\sum_{k=1}^{3} \frac{\partial g_i}{\partial \alpha_k}\frac{U_k}{\sqrt{(g_k)}} \quad (61a)$$

$$\gamma_{ij} = \frac{1}{\sqrt{(g_i g_j)}}\left[g_i \frac{\partial}{\partial \alpha_j}\left(\frac{U_i}{\sqrt{(g_i)}}\right) + g_j \frac{\partial}{\partial \alpha_i}\left(\frac{U_j}{\sqrt{(g_j)}}\right)\right] \quad (61b)$$

where $i, j = 1, 2, 3$ and the only sum is indicated explicitly. The quantities g_i are related to *metrics*, and in order to use our shell coordinates, we note the following correspondence:

$$\begin{aligned}
\alpha_1 &= \alpha_1 & \alpha_2 &= \alpha_2 & \alpha_3 &= z \\
U_1 &= U_1 & U_2 &= U_2 & U_3 &= W \\
g_1 &= A_1^2(1+z/R_1)^2 & g_2 &= A_2^2(1+z/R_2)^2 & g_3 &= 1.
\end{aligned} \quad (62)$$

Then it follows that

$$\begin{aligned}
\varepsilon_1 &= \frac{1}{A_1(1+z/R_1)}\left(\frac{\partial U_1}{\partial \alpha_1} + \frac{U_2}{A_2}\frac{\partial A_1}{\partial \alpha_2} + \frac{A_1 W}{R_1}\right) \\
\varepsilon_2 &= \frac{1}{A_2(1+z/R_2)}\left(\frac{\partial U_2}{\partial \alpha_2} + \frac{U_1}{A_1}\frac{\partial A_2}{\partial \alpha_1} + \frac{A_2 W}{R_2}\right) \\
\varepsilon_n &= \partial W/\partial z \\
\gamma_{12} &= \frac{A_2(1+z/R_2)}{A_1(1+z/R_1)}\frac{\partial}{\partial \alpha_1}\left[\frac{U_2}{A_2(1+z/R_2)}\right] \\
&\quad + \frac{A_1(1+z/R_1)}{A_2(1+z/R_2)}\frac{\partial}{\partial \alpha_2}\left[\frac{U_1}{A_1(1+z/R_1)}\right]
\end{aligned} \quad (63)$$

[3] See Appendix IIIB for a detailed derivation.

$$\gamma_{1n} = \frac{1}{A_1(1+z/R_1)}\frac{\partial W}{\partial \alpha_1} + A_1(1+z/R_1)\frac{\partial}{\partial z}\left[\frac{U_1}{A_1(1+z/R_1)}\right]$$

$$\gamma_{2n} = \frac{1}{A_2(1+z/R_2)}\frac{\partial W}{\partial \alpha_2} + A_2(1+z/R_2)\frac{\partial}{\partial z}\left[\frac{U_2}{A_2(1+z/R_2)}\right]. \quad (63)$$

Note that within the framework of infinitesimal elasticity, these relations are *exact*, i.e., no approximations have been made at all.

We now introduce the following *displacement distribution assumptions*:

$$\begin{aligned} U_1(\alpha_1, \alpha_2, z) &= u_1(\alpha_1, \alpha_2) + z\beta_1(\alpha_1, \alpha_2) \\ U_2(\alpha_1, \alpha_2, z) &= u_2(\alpha_1, \alpha_2) + z\beta_2(\alpha_1, \alpha_2) \\ W(\alpha_1, \alpha_2, z) &= w(\alpha_1, \alpha_2). \end{aligned} \quad (64)$$

Equations (64) may be viewed as the first terms in Taylor series expansions through the wall thickness coordinate, truncated early due to the fact that we are dealing only with thin shells. It can also be seen immediately that $\varepsilon_n \equiv 0$, due to the last of equations (64). The quantities β_1, β_2 clearly represent rotations of tangents to the reference surface, along the coordinate directions α_1, α_2, and they will be determined by requiring $\gamma_{1n} = \gamma_{2n} = 0$. Substituting for U_1, U_2, W from equations (64) into the last of equations (63) yields:

$$\begin{aligned} \gamma_{1n} &= \frac{1}{A_1(1+z/R_1)}\frac{\partial w}{\partial \alpha_1} \\ &\quad + A_1(1+z/R_1)\frac{\partial}{\partial z}\left[\frac{U_1}{A_1(1+z/R_1)} + \frac{z\beta_1}{A_1(1+z/R_1)}\right] \\ &= \frac{1}{A_1(1+z/R_1)}\left\{\frac{\partial w}{\partial \alpha_1} - \frac{A_1 u_1}{R_1} + \beta_1 A_1\right\} = 0 \end{aligned}$$

so that we obtain (together with an analogous calculation from $\gamma_{2n} = 0$)

$$\begin{aligned} \beta_1 &= \frac{u_1}{R_1} - \frac{1}{A_1}\frac{\partial w}{\partial \alpha_1} \\ \beta_2 &= \frac{u_2}{R_2} - \frac{1}{A_2}\frac{\partial w}{\partial \alpha_2}. \end{aligned} \quad (65)$$

Substitution of the distributions (64) into the exact strain–displacement relations yields

$$\varepsilon_1 = \frac{1}{(1+z/R_1)}(\varepsilon_1^0 + z\kappa_1) \tag{66a}$$

$$\varepsilon_2 = \frac{1}{(1+z/R_2)}(\varepsilon_2^0 + z\kappa_2) \tag{66b}$$

$$\gamma_{12} = \frac{1}{(1+z/R_1)}(\omega_1 + z\tau_1) + \frac{1}{(1+z/R_2)}(\omega_2 + z\tau_2) \tag{66c}$$

$$\varepsilon_n = \gamma_{1n} = \gamma_{2n} = 0 \tag{66d}$$

where

$$\varepsilon_1^0 = \frac{1}{A_1}\frac{\partial u_1}{\partial \alpha_1} + \frac{u_2}{A_1 A_2}\frac{\partial A_1}{\partial \alpha_2} + \frac{w}{R_1} \tag{66e}$$

$$\varepsilon_2^0 = \frac{1}{A_2}\frac{\partial u_2}{\partial \alpha_2} + \frac{u_1}{A_1 A_2}\frac{\partial A_2}{\partial \alpha_1} + \frac{w}{R_2} \tag{66f}$$

$$\kappa_1 = \frac{1}{A_1}\frac{\partial \beta_1}{\partial \alpha_1} + \frac{\beta_2}{A_1 A_2}\frac{\partial A_1}{\partial \alpha_2} \tag{66g}$$

$$\kappa_2 = \frac{1}{A_2}\frac{\partial \beta_2}{\partial \alpha_2} + \frac{\beta_1}{A_1 A_2}\frac{\partial A_2}{\partial \alpha_1} \tag{66h}$$

$$\omega_1 = \frac{1}{A_1}\frac{\partial u_2}{\partial \alpha_1} - \frac{u_1}{A_1 A_2}\frac{\partial A_1}{\partial \alpha_2} \tag{66i}$$

$$\omega_2 = \frac{1}{A_2}\frac{\partial u_1}{\partial \alpha_2} - \frac{u_2}{A_1 A_2}\frac{\partial A_2}{\partial \alpha_1} \tag{66j}$$

$$\tau_1 = \frac{1}{A_1}\frac{\partial \beta_2}{\partial \alpha_1} - \frac{\beta_1}{A_1 A_2}\frac{\partial A_1}{\partial \alpha_2} \tag{66k}$$

$$\tau_2 = \frac{1}{A_2}\frac{\partial \beta_1}{\partial \alpha_2} - \frac{\beta_2}{A_1 A_2}\frac{\partial A_2}{\partial \alpha_1}. \tag{66l}$$

In addition to the above, introduce the further notation

$$\omega = \omega_1 + \omega_2$$
$$\tau^* = \tau_1 + \frac{\omega_2}{R_1} \equiv \tau_2 + \frac{\omega_1}{R_2}. \tag{67}$$

The verification of the second half of the second of equations (67) is displayed in Appendix IIIC. It requires substitution from equations (65), (66) and the Codazzi conditions (28). With this notation it is easy to show that equation (66c) can also be written as

$$\gamma_{12} = \frac{1}{(1+z/R_1)(1+z/R_2)}\left[\omega(1-z^2/R_1R_2) + 2z\tau^*\left(1+\frac{z}{2}\left(\frac{1}{R_1}+\frac{1}{R_2}\right)\right)\right]. \tag{68}$$

III-4. STRESS RESULTANTS AND STRAIN ENERGY

We will now define a system of *stress resultants*, as follows:

$$[N_1, M_1] = \int_{-h/2}^{h/2} \sigma_1(1+z/R_2)[1, z]dz \tag{69a}$$

$$[N_2, M_2] = \int_{-h/2}^{h/2} \sigma_2(1+z/R_1)[1, z]dz \tag{69b}$$

$$[N_{12}, M_{12}] = \int_{-h/2}^{h/2} \tau_{12}(1+z/R_2)[1, z]dz \tag{69c}$$

$$[N_{21}, M_{21}] = \int_{-h/2}^{h/2} \tau_{21}(1+z/R_1)[1, z]dz. \tag{69d}$$

Note immediately that, although $\tau_{12} = \tau_{21}$, from equations (69) we see that $N_{12} \neq N_{21}$ and $M_{21} \neq M_{12}$. We shall later also define some shear force resultants Q_1, Q_2, which we do not yet need.

Various motivations may be given for the definitions (69), one of which comes simply from the *trapezoidal areas* displayed in Fig. 4 (p. 24), the other coming from an energy viewpoint to be elaborated below. As far as Fig. 4 and the associated two-dimensional stress field are concerned, we are simply summing all elements of area $d\Sigma(z)$ with the appropriate stress, to get the resultant:

$$A_2(d\alpha_2)N_1 = \sum_{\substack{\text{sum over}\\ \text{all } dz}} \sigma_1 ds_2(z) = \sum_{\substack{\text{sum over}\\ \text{all } dz}} \sigma_1(1+z/R_2)d\alpha_2 A_2$$

$$= \int_{-h/2}^{h/2} \sigma_1(1+z/R_2)dzd\alpha_2 A_2$$

where $A_2 d\alpha_2$ appears on the left since N_1 is a force per unit length. Similarly,

$$A_2(d\alpha_2)M_1 = \sum_{\substack{\text{sum over}\\ \text{all } dz}} (z)\sigma_1 ds_2(z) = \sum_{\substack{\text{sum over}\\ \text{all } dz}} z\sigma_1(1+z/R_2)d\alpha_2 A_2$$

$$= \int_{-h/2}^{h/2} \sigma_1 z(1+z/R_2)dz(d\alpha_2)A_2.$$

Now we switch to a consideration of the strain energy in a shell. By definition

$$U_e = \tfrac{1}{2}\int_V (\sigma_1\varepsilon_1+\sigma_2\varepsilon_2+\tau_{12}\gamma_{12})dV. \tag{70}$$

It is not difficult to show that

$$\delta U_e = \int_V (\sigma_1\delta\varepsilon_1+\sigma_2\delta\varepsilon_2+\tau_{12}\delta\gamma_{12})dV \tag{71}$$

as a consequence of the linearity of the constitutive law implied by equation (70).

We recognize equation (71) as a virtual work statement, and therefore independent of the nature of the material! One should really start with equation (71) and then assume a constitutive law!!

In any event, we will proceed from equation (71). For the volume element, from Fig. 4,

$$dV = A_1(1+z/R_1)A_2(1+z/R_2)d\alpha_1 d\alpha_2 dz \tag{72}$$

and then subsituting from equations (66) we find

$$\delta U_e = \iiint_{\text{surf. } z} \Big[\left(\frac{\sigma_1}{1+z/R_1}\right)(\delta\varepsilon_1^0+z\delta\kappa_1)+\left(\frac{\sigma_2}{1+z/R_2}\right)(\delta\varepsilon_2^0+z\delta\kappa_2)$$

$$+\left(\frac{\tau_{12}}{1+z/R_1}\right)(\delta\omega_1+z\delta\tau_1)+\left(\frac{\tau_{12}}{1+z/R_2}\right)(\delta\omega_2+z\delta\tau_2)\Big]dV. \tag{73}$$

If we substitute for the volume element (72), we can then write U_e in an expanded form, where the stress resultant definitions (69) appear very naturally. If we do this we find

$$\delta U_e = \iint (N_1\delta\varepsilon_1^0+M_1\delta\kappa_1+N_2\delta\varepsilon_2^0+M_2\delta\kappa_2+N_{12}\delta\omega_1 + N_{21}\delta\omega_2+M_{12}\delta\tau_1+M_{21}\delta\tau_2)A_1A_2d\alpha_1d\alpha_2. \tag{74}$$

We now define two new stress variables,

$$S = N_{12}-\frac{M_{21}}{R_2} = N_{21}-\frac{M_{12}}{R_1} \tag{75a}$$

$$H = \tfrac{1}{2}(M_{12}+M_{21}). \tag{75b}$$

The right half of equations (75a) is easily verified from the definitions (69). In fact we also note that (75a) can be written in the form

$$N_{12}-N_{21}+\frac{M_{12}}{R_1}-\frac{M_{21}}{R_2} = 0. \tag{75a$'$}$$

Substitution of equations (75, 67) into equation (74) yields:

$$\delta U_e = \iint (N_1\delta\varepsilon_1^0+M_1\delta\kappa_1+N_2\delta\varepsilon_2^0+M_2\delta\kappa_2 + S\delta\omega+2H\delta\tau^*)A_1A_2d\alpha_1d\alpha_2. \tag{76}$$

Thus the strain energy due to the shell deformation is now reduced to a surface integral involving the energy of stretching and bending the surface.

III-5. EQUATIONS OF EQUILIBRIUM

To formulate a complete set of equilibrium equations, one should write down a complete Lagrangian

$$L = T-(U_e+V)$$

where T = kinetic energy, U_e = strain energy, V = potential of applied edge and surface loads, and then properly apply Hamilton's principle,

$$\delta\int_{t_1}^{t_2} L\,dt = 0.$$

For ease and clarity at this stage we will ignore the kinetic energy (thus restricting ourselves to static problems) and include only a normal surface loading, i.e.,

$$V = +\iint q_n(\alpha_1, \alpha_2)w(\alpha_1, \alpha_2)A_1A_2 d\alpha_1 d\alpha_2 \tag{77}$$

so that

$$\delta V = +\iint q_n \delta w A_1 A_2 d\alpha_1 d\alpha_2 . \tag{77a}$$

Now by use of the principle of minimum potential energy, we see that

$$\delta(U_e+V) = \iint (N_1\delta\varepsilon_1^0 + N_2\delta\varepsilon_2^0 + S\delta\omega + M_1\delta\kappa_1 + M_2\delta\kappa_2 + 2H\delta\tau^* + q_n\delta w)A_1A_2 d\alpha_1 d\alpha_2 . \tag{78}$$

Now by equations (66) we see that ε_1^0, ε_2^0, κ_1, κ_2, ω and τ^* can be written in terms of u_1, u_2, w, β_1 and β_2. But, by equations (65), β_1 and β_2 are functions of u_1, u_2 and w. Thus we have here two choices: (1) We can proceed with the variation (78), using equations (65, 66) with independent variations only on u_1, u_2 and w. This will give us a "reduced" system of three equilibrium equations. (2) On the other hand, we could allow for "non-vanishing strain variations $\delta\gamma_{1n}$, $\delta\gamma_{2n}$", even though we assume $\gamma_{1n} = \gamma_{2n} = 0$, to allow for the strain energy due to the internal work of the shear forces. Thus we would define shear force resultants

$$\begin{aligned} Q_1 &= \int_{-h/2}^{h/2} \tau_{1n}(1+z/R_2)dz \\ Q_2 &= \int_{-h/2}^{h/2} \tau_{2n}(1+z/R_1)dz . \end{aligned} \tag{79}$$

Then, the term that we would add to the strain energy variation is

$$\delta U_e\big|_{\text{add}} = \iiint (\tau_{1n}\delta\gamma_{1n} + \tau_{2n}\delta\gamma_{2n})A_1(1+z/R_1)A_2(1+z/R_2)d\alpha_1 d\alpha_2 dz .$$

From p. 26 [just prior to equation (65)],

$$\delta\gamma_{1n} = \frac{1}{A_1(1+z/R_1)}\left\{\frac{\partial\delta w}{\partial\alpha_1} - \frac{A_1}{R_1}\delta u_1 + A_1\delta\beta_1\right\}$$

$$\delta\gamma_{2n} = \frac{1}{A_2(1+z/R_2)}\left\{\frac{\partial\delta w}{\delta\alpha_2} - \frac{A_2}{R_2}\delta u_2 + A_2\delta\beta_2\right\}$$

so that in view of the above, and the definitions (79) the additional energy variation terms are

$$\delta U_e \Big|_{\text{add}} = \iint\limits_{\text{surf.}} \Big[Q_1 A_2 \Big\{ \frac{\partial \delta w}{\partial \alpha_1} - \frac{A_1}{R_1} \delta u_1 + A_1 \delta \beta_1 \Big\}$$

$$+ Q_2 A_1 \Big\{ \frac{\partial \delta w}{\partial \alpha_2} - \frac{A_2}{R_2} \delta u_2 + A_2 \delta \beta_2 \Big\} \Big] d\alpha_1 d\alpha_2. \tag{80}$$

We shall also find it convenient, when going through this second approach, to leave the shear forces and twisting moments in their original untransformed state, i.e., as in equation (74). *Thus we shall vary*

$$\delta U_e = \iint \Big[N_1 \Big\{ A_2 \frac{\partial \delta u_1}{\partial \alpha_1} + \frac{\partial A_1}{\partial \alpha_2} \delta u_2 + A_1 A_2 \frac{\delta w}{R_1} \Big\}$$

$$+ M_1 \Big\{ A_2 \frac{\partial \delta \beta_1}{\partial \alpha_1} + \frac{\partial A_1}{\partial \alpha_2} \delta \beta_2 \Big\} + N_2 \Big\{ A_1 \frac{\partial \delta u_2}{\delta \alpha_2}$$

$$+ \frac{\partial A_2}{\partial \alpha_1} \delta u_1 + A_1 A_2 \frac{\delta w}{R_2} \Big\} + M_2 \Big\{ A_1 \frac{\partial \delta \beta_2}{\partial \alpha_2} + \frac{\partial A_2}{\partial \alpha_1} \delta \beta_1 \Big\}$$

$$+ N_{12} \Big\{ A_2 \frac{\partial \delta u_2}{\partial \alpha_1} - \frac{\partial A_1}{\partial \alpha_2} \delta u_1 \Big\} + N_{21} \Big\{ A_1 \frac{\partial \delta u_1}{\partial \alpha_2} - \frac{\partial A_2}{\partial \alpha_1} \delta u_2 \Big\}$$

$$+ M_{12} \Big\{ A_2 \frac{\partial \delta \beta_2}{\partial \alpha_1} - \frac{\partial A_1}{\partial \alpha_2} \delta \beta_1 \Big\} + M_{21} \Big\{ A_1 \frac{\partial \delta \beta_1}{\partial \alpha_2} - \frac{\partial A_2}{\partial \alpha_1} \delta \beta_2 \Big\}$$

$$+ Q_1 \Big\{ A_2 \frac{\partial \delta w}{\partial \alpha_1} + A_1 A_2 \Big(\delta \beta_1 - \frac{\delta u_1}{R_1} \Big) \Big\}$$

$$+ Q_2 \Big\{ A_1 \frac{\partial \delta w}{\partial \alpha_2} + A_1 A_2 \Big(\delta \beta_2 - \frac{\delta u_2}{R_2} \Big) \Big\} \Big] d\alpha_1 d\alpha_2 . \tag{81}$$

Then variation of equations (77a), (81) yields the following *equations of equilibrium:*

$$\delta u_1: \quad \frac{\partial (N_1 A_2)}{\partial \alpha_1} + \frac{\partial (N_{21} A_1)}{\partial \alpha_2} + N_{12} \frac{\partial A_1}{\partial \alpha_2} - N_2 \frac{\partial A_2}{\partial \alpha_1} + A_1 A_2 \frac{Q_1}{R_1} = 0 \tag{82a}$$

$$\delta u_2: \quad \frac{\partial(N_{12}A_2)}{\partial\alpha_1}+\frac{\partial(N_2A_1)}{\partial\alpha_2}+N_{21}\frac{\partial A_2}{\partial\alpha_1}-N_1\frac{\partial A_1}{\partial\alpha_2}+A_1A_2\frac{Q_2}{R_2}=0 \quad (82b)$$

$$\delta w: \quad \frac{\partial(Q_1A_2)}{\partial\alpha_1}+\frac{\partial(Q_2A_1)}{\partial\alpha_2}-\left(\frac{N_1}{R_1}+\frac{N_2}{R_2}\right)A_1A_2=q_nA_1A_2 \quad (82c)$$

$$\delta\beta_1: \quad \frac{\partial(M_1A_2)}{\partial\alpha_1}+\frac{\partial(M_{21}A_1)}{\partial\alpha_2}+M_{12}\frac{\partial A_1}{\partial\alpha_2}-M_2\frac{\partial A_2}{\partial\alpha_1}$$
$$-Q_1A_2A_2=0 \quad (82d)$$

$$\delta\beta_2: \quad \frac{\partial(M_{12}A_2)}{\partial\alpha_1}+\frac{\partial(M_2A_1)}{\partial\alpha_2}+M_{21}\frac{\partial A_2}{\partial\alpha_1}-M_1\frac{\partial A_1}{\partial\alpha_2}$$
$$-Q_2A_1A_2=0. \quad (82e)$$

Note that we have above five differential equations in the ten variables $N_1, N_2, N_{12}, N_{21}, M_1, M_2, M_{12}, M_{21}, Q_1$ and Q_2. We shall shortly show that equations (69), (66), and (52), (54) can be combined to yield eight more equations relating the first eight stress resultants to the five kinematic quantities u_1, u_2, w, β_1 and β_2, of which only three are independent due to equations (65).

The *boundary conditions* resulting from the variation of equations (77a), (81) corresponding to the differential equations (82) are:

At constant α_1: $N_1=0$ or $\delta u_1=0$

$N_{12}+M_{12}/R_2=0$ or $\delta u_2=0$

$Q_1+\partial M_{12}/A_2\partial\alpha_2=0$ or $\delta w=0$

$M_1=0$ or $\delta\beta_1=0$. (83)

At constant α_2: $N_2=0$ or $\delta u_2=0$

$N_{21}+M_{21}/R_1=0$ or $\delta u_1=0$

$Q_2+\partial M_{21}/A_1\partial\alpha_1=0$ or $\delta w=0$

$M_2=0$ or $\delta\beta_2=0$.

We defer discussion of these boundary conditions—particularly the *Kirchoff effective shears*—at this time. To complete the present picture we present now the *force-displacement* relations obtained by substituting equations (66), (52), (54) into equations (69):

$$N_1 = \frac{E}{1-\nu^2}\int_{-h/2}^{h/2}\left[\left(\frac{1+z/R_2}{1+z/R_1}\right)(\varepsilon_1^0+z\kappa_1)+\nu(\varepsilon_2^0+z\kappa_2)\right]dz \tag{84a}$$

$$N_{12} = \frac{E}{2(1+\nu)}\int_{-h/2}^{h/2}\left\{(1-z^2/R_1R_2)\omega + 2z\tau^*\left[1+\frac{z}{2}(1/R_1+1/R_2)\right]\right\}\frac{dz}{(1+z/R_1)} \tag{84b}$$

$$M_1 = \frac{E}{1-\nu^2}\int_{-h/2}^{h/2}\left[\left(\frac{1+z/R_2}{1+z/R_1}\right)(\varepsilon_1^0+z\kappa_1)+\nu(\varepsilon_2^0+z\kappa_2)\right]zdz \tag{84c}$$

$$M_{12} = \frac{E}{2(1+\nu)}\int_{-h/2}^{h/2}\left\{(1-z^2/R_1R_2)\omega + 2z\tau^*\left[1+\frac{z}{2}(1/R_1+1/R_2)\right]\right\}\frac{zdz}{(1+z/R_1)} \tag{84d}$$

and four more equations[4] for N_2, N_{21}, M_2, and M_{21} obtained from the above by interchanging subscripts $1 \to 2$ and $2 \to 1$.

These sets of equations (82), (83), (84), (66) complete in a sense a formulation of a linear shell theory. Only the plane stress assumption and the hairbrush hypothesis have been used—we have made no approximations on the basis of $h/R_i \ll 1$ or $z/R_i \ll 1$.

Before moving off the formulation phase, we note that in accordance with the discussion on p. 31 the following alternative (1) is available, i.e., to use the variational formulation (81) with $Q_1 \equiv Q_2 \equiv 0$, and with β_1, β_2 eliminated throughout by substitution of equations (65). Thus we would get three equations of equilibrium (δu_1, δu_2, and δw).

[4] Referred to as (84e, f, g, h) even though not listed.

Then, in addition, we solve equations (82d, e) for Q_1 and Q_2 and substitute back into equations (82a, b, c), thus eliminating Q_1, Q_2. The three equations should be identical to those three derived variationally as just described. These results are indicated in Appendix IIID.

III-6. SIMPLICATIONS OF THE STRAIN ENERGY FUNCTIONAL AND THE STRESS–STRAIN RELATIONS

As a first consideration towards simplifying approximations, let us examine the strain energy in the form [substituting from equations (52), (54) into equations (70)]:

$$U_e = \frac{1}{2}\int_V \left[\frac{E}{1-\nu^2}(\varepsilon_1^2+2\nu\varepsilon_1\varepsilon_2+\varepsilon_2^2)+G\gamma_{12}^2\right]dV$$

$$= \frac{E}{2(1-\nu^2)}\int_V \left[\varepsilon_1^2+2\nu\varepsilon_1\varepsilon_2+\varepsilon_2^2+\frac{1-\nu}{2}\gamma_{12}^2\right]dV$$

$$= \frac{E}{2(1-\nu^2)}\int_V \left[(\varepsilon_1+\varepsilon_2)^2-2(1-\nu)\left(\varepsilon_1\varepsilon_2-\frac{1}{4}\gamma_{12}^2\right)\right]dV. \tag{85}$$

Now we will substitute from equations (66), (68), (72), using the shorthand notation that $dS = A_1A_2d\alpha_1d\alpha_2$ and

$$\varepsilon_i' = \frac{h}{2}\kappa_i, \quad \tau' = h\tau^*. \tag{86}$$

Equations (86) serve the purpose of defining ε_i' and τ' as strains in the outer fibres due to bending and twisting, respectively. They also imply that the ε_i' will be of the same order of magnitude as ε_i^0, and that the τ' will be of the same order as ω. Then

$$U_e = \frac{E}{2(1-\nu^2)}\iiint\left\{\left(\frac{\varepsilon_1^0+(2z/h)\varepsilon_1'}{1+z/R_1}+\frac{\varepsilon_2^0+(2z/h)\varepsilon_2'}{1+z/R_2}\right)^2\right.$$

$$-2(1-\nu)\left(\frac{\varepsilon_1^0+(2z/h)\varepsilon_1'}{1+z/R_1}\right)\left(\frac{\varepsilon_2^0+(2z/h)\varepsilon_2'}{1+z/R_2}\right)$$

$$\left.+\frac{1-\nu}{2}\left(\frac{\omega(1-z^2/R_1R_2)+(2z/h)\tau'[1+(z/2R_1)+(z/2R_2)]}{(1+z/R_1)(1+z/R_2)}\right)^2\right\}\times$$

$$\times(1+z/R_1)(1+z/R_2)dzdS$$

$$U_e = \frac{E}{2(1-\nu^2)} \iint \int_{-h/2}^{h/2} \left\{ \left(\varepsilon_1^0 + \frac{2z}{h} \varepsilon_1' \right)^2 \frac{1+z/R_2}{1+z/R_1} + 2\left(\varepsilon_1^0 + \frac{2z}{h} \varepsilon_1' \right) \times \right.$$

$$\times \left(\varepsilon_2^0 + \frac{2z}{h} \varepsilon_2' \right) + \left(\varepsilon_2^0 + \frac{2z}{h} \varepsilon_2' \right)^2 \frac{1+z/R_1}{1+z/R_2} - 2(1-\nu)\left(\varepsilon_1^0 + \frac{2z}{h} \varepsilon_1' \right) \times$$

$$\times \left(\varepsilon_2^0 + \frac{2z}{h} \varepsilon_2' \right) + \frac{1-\nu}{2} \left[\omega(1 - z^2/R_1R_2) \right.$$

$$\left. \left. + \frac{2z}{h} \tau'(1+z/2R_1+z/2R_2) \right]^2 (1+z/R_1)^{-1}(1+z/R_2)^{-1} \right\} dz dS .$$

If we then expand $(1+z/R_i)^{-1} \cong 1 - z/R_i + (z/R_i)^2$ so that we keep the integrand accurate up to terms of $0(z^2)$ we find that

$$U_e = \frac{E}{2(1-\nu^2)} \iint \int_{-h/2}^{h/2} (X_0 + X_1 z + X_2 z^2) dz dS \tag{87a}$$

where

$$X_0 = (\varepsilon_1^0 + \varepsilon_2^0)^2 - 2(1-\nu)\left(\varepsilon_1^0 \varepsilon_2^0 - \frac{\omega^2}{4} \right) \tag{87b}$$

$$X_2 = \frac{4}{h^2} \left[(\varepsilon_1' + \varepsilon_2')^2 - 2(1-\nu)\left(\varepsilon_1' \varepsilon_2' - \frac{\tau'^2}{4} \right) \right]$$

$$+ \frac{4}{h} \left[\left(\frac{1}{R_2} - \frac{1}{R_1} \right)(\varepsilon_1^0 \varepsilon_1' - \varepsilon_2^0 \varepsilon_2') - \left(\frac{1-\nu}{4} \right)\left(\frac{1}{R_1} + \frac{1}{R_2} \right) \omega \tau' \right]$$

$$+ \left[\left(\frac{1}{R_2} - \frac{1}{R_1} \right)\left(\frac{\varepsilon_2^{0 2}}{R_2} - \frac{\varepsilon_1^{0 2}}{R_1} \right) + \omega^2 \left(\frac{1}{R_1^2} - \frac{1}{R_1 R_2} + \frac{1}{R_2^2} \right)\left(\frac{1-\nu}{2} \right) \right] \tag{87c}$$

where the X_1 term does not matter because we assume that

$$\int_{-h/2}^{h/2} (X_0 + X_1 z + X_2 z^2) dz = hX_0 + \frac{h^3}{12} X_2 \tag{88}$$

and thus the X_1 term does not enter into the final form of the strain energy. Then after substitution of the result (88) we find that

$$U_e = \frac{Eh}{2(1-\nu^2)} \iint \left\{ \left[(\varepsilon_1^0 + \varepsilon_2^0)^2 + 2(1-\nu)\left(\varepsilon_1^0 \varepsilon_2^0 - \frac{\omega^2}{4} \right) \right] \right.$$

$$+ \frac{1}{3}\left[(\varepsilon_1' + \varepsilon_2')^2 - 2(1-\nu)\left(\varepsilon_1' \varepsilon_2' - \frac{\tau'^2}{4} \right) \right]$$

$$+ \frac{1}{3}\left[\left(\frac{h}{R_2} - \frac{h}{R_1} \right)(\varepsilon_1^0 \varepsilon_1' - \varepsilon_2^0 \varepsilon_2') - \left(\frac{1-\nu}{4} \right)\left(\frac{h}{R_1} + \frac{h}{R_2} \right)\omega\tau' \right]$$

$$+ \frac{1}{12}\left[\left(\frac{h}{R_2} - \frac{h}{R_1} \right)\left(\frac{h}{R_2}\, \varepsilon_2^{0\,2} - \frac{h}{R_1}\, \varepsilon_1^{0\,2} \right) \right.$$

$$\left. \left. + \frac{1-\nu}{2}\left(\frac{h^2}{R_1^2} - \frac{h^2}{R_1 R_2} + \frac{h^2}{R_2^2} \right)\omega^2 \right] \right\} dS. \tag{89}$$

Recall that due to the transformations (86) all the terms ε_i^0, ε_i', τ' and ω are of the same order of magnitude. Thus we can note that the integrand in equation (89) has terms of $0(1)$, $0(h/R_i)$, and $0[(h/R_i)^2]$. *If we now restrict ourselves to thin shells, for which we can neglect h/R_i [and thus certainly $(h/R_i)^2$] compared to unity, then the thin shell strain energy can be taken as*

$$U_e = \frac{Eh}{2(1-\nu^2)} \iint \left\{ \left[(\varepsilon_1^0 + \varepsilon_2^0)^2 + 2(1-\nu)\left(\varepsilon_1^0 \varepsilon_2^0 - \frac{\omega^2}{4} \right) \right] \right.$$

$$\left. + \frac{1}{3}\left[(\varepsilon_1' + \varepsilon_2')^2 - 2(1-\nu)\left(\varepsilon_1' \varepsilon_2' - \frac{\tau'^2}{4} \right) \right] \right\} A_1 A_2 d\alpha_1 d\alpha_2 . \tag{90}$$

Reverting now from the notation of equation (86),

$$U_e = \frac{1}{2} C \iint \left[(\varepsilon_1^0 + \varepsilon_2^0)^2 - 2(1-\nu)\left(\varepsilon_1^0 \varepsilon_2^0 - \frac{\omega^2}{4} \right) \right] A_1 A_2 d\alpha_1 d\alpha_2$$

$$+ \frac{1}{2} D \iint [(\kappa_1 + \kappa_2)^2 - 2(1-\nu)(\kappa_1 \kappa_2 - \tau^{*2})] A_1 A_2 d\alpha_1 d\alpha_2 \tag{91}$$

where we have introduced the *extensional rigidity* C and *bending rigidity* D, defined by

$$C = \frac{Eh}{1-\nu^2}, \quad D = \frac{Eh^3}{12(1-\nu^2)}. \tag{92}$$

To define appropriate relations between the stress resultants N_1, N_2, M_1, M_2, S and H, we will now compare the variation (78) to the variation of the energy (91). The latter is easily found to be

$$\delta U_e = \iint \Big[C(\varepsilon_1^0 + \nu\varepsilon_2^0)\delta\varepsilon_1^0 + C(\varepsilon_2^0 + \nu\varepsilon_1^0)\delta\varepsilon_2^0$$
$$+ \frac{1-\nu}{2} C\omega\delta\omega + D(\kappa_1 + \nu\kappa_2)\delta\kappa_1 + D(\kappa_2 + \nu\kappa_1)\delta\kappa_2$$
$$+ 2(1-\nu)D\tau^*\delta\tau^* \Big] A_1 A_2 d\alpha_1 d\alpha_2 . \tag{93}$$

Then comparing equations (78), (93), we see that

$$\begin{aligned} N_1 &= C(\varepsilon_1^0 + \nu\varepsilon_2^0), & N_2 &= C(\varepsilon_2^0 + \nu\varepsilon_1^0) \\ M_1 &= D(\kappa_1 + \nu\kappa_2), & M_2 &= D(\kappa_2 + \nu\kappa_1) \\ S &= \left(\frac{1-\nu}{2}\right) C\omega, & H &= (1-\nu)D\tau^* . \end{aligned} \tag{94}$$

In view of equations (75) we can write

$$\begin{aligned} N_{12} - \frac{M_{21}}{R_2} &= \frac{Eh}{2(1+\nu)}\,\omega \\ N_{21} - \frac{M_{12}}{R_1} &= \frac{Eh}{2(1+\nu)}\,\omega \\ M_{12} + M_{21} &= \frac{Eh^3}{6(1+\nu)}\,\tau^* \end{aligned} \tag{95}$$

so that simply knowing ω, τ^* is not sufficient to determine N_{12}, N_{21}, M_{12}, M_{21} uniquely. In fact one can write that

$$N_{12} = \frac{Eh}{2(1+\nu)}\left(\omega + \frac{h^2}{6R_2}\tau^*\right) - \Phi/R_2$$

$$N_{21} = \frac{Eh}{2(1+\nu)}\left(\omega + \frac{h^2}{6R_1}\tau^*\right) + \Phi/R_1$$

$$M_{12} = \frac{Eh^3}{12(1+\nu)}\tau^* + \Phi \tag{96}$$

$$M_{21} = \frac{Eh^3}{12(1+\nu)}\tau^* - \Phi$$

where the function Φ can be chosen for any convenient purpose, and it plays no role of importance in the development of further shell theories. Now from the last two of equations (96) and the definitions (69c, d)

$$M_{12} - M_{21} = 2\Phi = \left(\frac{1}{R_2} - \frac{1}{R_1}\right)\int_{-h/2}^{h/2} \tau_{12} z^2 dz\,. \tag{97}$$

From equations (54), (68) we can substitute for τ_{12} above to write

$$M_{12} - M_{21} = 2\Phi = \left(\frac{1}{R_2} - \frac{1}{R_1}\right)\int_{-h/2}^{h/2} G\gamma_{12} z^2 dz$$

or

$$\Phi = \frac{G}{2}\left(\frac{1}{R_2} - \frac{1}{R_1}\right)\int_{-h/2}^{h/2}\left\{\frac{z^2}{(1+z/R_1)(1+z/R_2)}\left[\omega(1 - z^2/R_1R_2) + 2z\tau^*\left(1 + \frac{z}{2R_1} + \frac{z}{2R_2}\right)\right]\right\}dz\,. \tag{98}$$

Now if we neglect z/R_i compared to unity,

$$\Phi \cong \left(\frac{1}{R_2} - \frac{1}{R_1}\right)\frac{Eh^3}{48(1+\nu)}\omega\,. \tag{99}$$

If we were to substitute equations (96) we would find first that

$$N_{12} = \frac{Eh}{2(1+\nu)}\left(\omega + \frac{h^2}{6R_2}\tau^*\right) - \frac{Eh}{2(1+\nu)}\left(\frac{h^2}{24}\right)\left(\frac{1}{R_1^2} + \frac{1}{R_2R_1}\right)\omega$$

$$\cong \frac{Eh}{2(1+\nu)}\left(\omega + \frac{h^2}{6R_2}\tau^*\right). \tag{100}$$

Also, using equations (96), (99), (86), we have

$$M_{12} = \frac{Eh^3}{12(1+\nu)}\left(\frac{\tau'}{h}\right) + \frac{Eh^3}{12(1+\nu)}\left(\frac{1}{R_2} - \frac{1}{R_1}\right)\frac{\omega}{4}$$

$$= \frac{Eh^2}{12(1+\nu)}\left[\tau' + \left(\frac{h}{R_2} - \frac{h}{R_1}\right)\frac{\omega}{4}\right] \cong \frac{Eh^2}{12(1+\nu)}\tau'$$

or

$$M_{12} = M_{21} \cong \frac{Eh^3}{12(1+\nu)}\tau^*$$

since we have established earlier that τ', ω are of the same order of magnitude.

Thus we may adopt the view (tentatively) that for thin shells, the stress–displacement relations can be taken as

$$\begin{aligned}
N_1 &= C(\varepsilon_1^0 + \nu\varepsilon_2^0), \quad N_2 = C(\varepsilon_2^0 + \nu\varepsilon_1^0) \\
M_1 &= D(\kappa_1 + \nu\kappa_2), \quad M_2 = D(\kappa_2 + \nu\kappa_1) \\
N_{12} &= \frac{Eh}{2(1+\nu)}\left(\omega + \frac{h^2}{6R_2}\tau^*\right) \\
N_{21} &= \frac{Eh}{2(1+\nu)}\left(\omega + \frac{h^2}{6R_1}\tau^*\right) \\
M_{12} &= M_{21} = \frac{Eh^3}{12(1+\nu)}\tau^*.
\end{aligned} \tag{101}$$

Now it is also clear, since $h\tau^* = \tau'$ and ω are of the same order of magnitude, that $\omega \gg (h^2\tau^*/6R_i)$, so that we might consider the possibility that $N_{12} \cong N_{21}$. In fact if we used the limiting process

$$1+z/R_1 \cong 1+z/R_2 \cong 1-z^2/R_1R_2 \cong 1$$

in the "exact" stress–displacement relations (84) we would find that

$$
\begin{aligned}
N_1 &= C(\varepsilon_1^0+\nu\varepsilon_2^0), \quad N_2 = C(\varepsilon_2^0+\nu\varepsilon_1^0)\\
N_{12} &= N_{21} = \left(\frac{1-\nu}{2}\right)C\omega = \frac{Eh}{2(1+\nu)}\,\omega\\
M_1 &= D(\kappa_1+\nu\kappa_2), \quad M_2 = D(\kappa_2+\nu\kappa_1)\\
M_{12} &= M_{21} = (1-\nu)D\tau^* = \frac{Eh^3}{12(1+\nu)}\tau^*.
\end{aligned}
\tag{102}
$$

It is not difficult to verify that equations (102) are *not* consistent variationally with the energy variations (91), (78), unless we identify $S = N_{12}$, $H = M_{12}$, which would appear to fit in with the assumptions leading to equations (102).

We now make one more (final) simplification, and thus we will obtain the formulation we desire for solving problems in the sequel. Note that using equations (67),

$$2\tau^* = \tau_1+\frac{\omega_2}{R_1}+\tau_2+\frac{\omega_1}{R_2} \equiv \tau+\frac{\omega_2}{R_1}+\frac{\omega_1}{R_2}$$

so that

$$2\tau^*h = \tau h+\omega_1\frac{h}{R_2}+\omega_2\frac{h}{R_1}.$$

We have earlier pointed out that $\tau^*h = \tau'$ is of the same order as ω, and thus of ω_1, ω_2. Clearly, then $2\tau^*h \cong \tau h$, and our *stress–displacement conditions* will be taken as

$$
\begin{aligned}
N_1 &= C(\varepsilon_1^0+\nu\varepsilon_2^0), \quad N_2 = C(\varepsilon_2^0+\nu\varepsilon_1^0)\\
N_{12} &= N_{21} = \left(\frac{1-\nu}{2}\right)C\omega = \frac{Eh}{2(1+\nu)}\,\omega\\
M_1 &= D(\kappa_1+\nu\kappa_2), \quad M_2 = D(\kappa_2+\nu\kappa_1)\\
M_{12} &= M_{21}\left(\frac{1-\nu}{2}\right)D\tau = \frac{Eh^3}{24(1+\nu)}\,\tau
\end{aligned}
\tag{103}
$$

where, again,

$$\tau \equiv \tau_1 + \tau_2. \tag{104}$$

This last equation, and the relations (103) for N_{12}, M_{12}, etc., are equivalent to taking

$$\gamma_{12} = \omega_1 + \omega_2 + z(\tau_1 + \tau_2) = \omega + z\tau. \tag{105}$$

If we go back to the variational equation (74) and allow $N_{12} = N_{21}$, and $M_{12} = M_{21}$, then

$$\begin{aligned} \delta U_e = \iint (&N_1 \delta \varepsilon_1^0 + M_1 \delta \kappa_1 + N_2 \delta \varepsilon_2^0 + M_2 \delta \kappa_2 \\ &+ N_{12} \delta \omega + M_{12} \delta \tau) A_1 A_2 d\alpha_1 d\alpha_2 . \end{aligned} \tag{106}$$

And if we replace τ^* by $\tau/2$ in equation (91), as clearly seems appropriate now,

$$\begin{aligned} U_e = \frac{1}{2} C \iint \left[(\varepsilon_1^0 + \varepsilon_2^0)^2 - 2(1-\nu)\left(\varepsilon_1^0 \varepsilon_2^0 - \frac{\omega^2}{4} \right) \right] A_1 A_2 d\alpha_1 d\alpha_2 \\ + \frac{1}{2} D \iint \left[(\kappa_1 + \kappa_2)^2 - 2(1-\nu)\left(\kappa_1 \kappa_2 - \frac{\tau^2}{4} \right) \right] A_1 A_2 d\alpha_1 d\alpha_2 . \end{aligned} \tag{107}$$

Then it is easily verified that δU_e (107) is such that, when equated to δU_e (106), the stress–strain relations for the shell are those given by equations (103). Thus they can be viewed as being variationally consistent. Furthermore, looking at the expanded version of equation (106) —i.e., letting $\omega = \omega_1 + \omega_2$ and $\tau = \tau_1 + \tau_2$—we see that the form is identical to that of equation (81), without the shear terms, or equation (74), except for the assumption $N_{12} = N_{21}$, $M_{12} = M_{21}$. Thus the equations of equilibrium and the boundary conditions corresponding to the vanishing of the variation (106 or 107) are given by equations (82), (83) with $N_{12} = N_{21}$ and $M_{12} = M_{21}$.

There is one drawback to the set of equations now proposed. One requirement of any set of equations for stresses and displacements in a structural element is that no stresses are created when the structural element undergoes a rigid body motion, i.e., a translation and rotation such that

$$\mathbf{U} = \mathbf{\Delta} + \mathbf{\Omega} \times \mathbf{R} \tag{108}$$

where

$$\begin{aligned} \boldsymbol{\Delta} &= \delta_1 \mathbf{t}_1 + \delta_2 \mathbf{t}_2 + \delta_n \mathbf{n} \\ \boldsymbol{\Omega} &= \Omega_2 \mathbf{t}_1 + \Omega_1 \mathbf{t}_2 + \Omega_n \mathbf{n} \\ \mathbf{R} &= \rho_1 \mathbf{t}_1 + \rho_2 \mathbf{t}_2 + \rho_n \mathbf{n} \,. \end{aligned} \tag{109}$$

It is shown in Appendix IIIE that for these displacements

$$\varepsilon_1^0 = \varepsilon_2^0 = 0, \quad \kappa_1 = \kappa_2 = 0, \quad \omega = 0, \quad \tau^* = 0$$

but that $\tau = \tau_1 + \tau_2 \neq 0$. Thus a rigid body motion, by equations (103), creates a non-zero moment resultant, the twisting moment M_{12}. This is due to the fact that by symmetrizing N_{12} and M_{12}, we can no longer satisfy equation (75a)′. However, in most practical problems, the error thus induced is small.

(In a classical paper of Sanders an alternate theory is derived which does have in it the vanishing of all stresses during rigid body motions.)

We also note that if we are interested in calculating the actual stresses in the shell, and not just the shell resultants, then

$$\begin{aligned} \sigma_1 &= \frac{E}{1-\nu^2}(\varepsilon_1 + \nu\varepsilon_2) = \frac{E}{1-\nu^2}[(\varepsilon_1^0 + \nu\varepsilon_2^0) + z(\kappa_1 + \nu\kappa_2)] \\ &= \frac{E}{1-\nu^2}\left[\frac{N_1}{C} + z\,\frac{M_1}{D}\right] = \frac{N_1}{h} + z\,\frac{M_1}{(h^3/12)}\,. \end{aligned} \tag{110a}$$

Similarly

$$\sigma_2 = \frac{N_2}{h} + z\,\frac{M_2}{(h^3/12)} \tag{110b}$$

$$\tau_{12} = \frac{N_{12}}{h} + z\,\frac{M_{12}}{(h^3/12)}\,. \tag{110c}$$

III-7. THE KIRCHOFF BOUNDARY CONDITIONS

As a final point of discussion, we consider the boundary conditions (83), in particular those of the form that at α_n = const.,

$$\begin{aligned} T_{nt} &\equiv N_{nt} + M_{nt}/R_t = 0 \quad \text{or} \quad \delta u_t = 0 \\ V_n &\equiv Q_n + \frac{\partial M_{nt}}{A_t \partial \alpha_t} = 0 \quad \text{or} \quad \delta w = 0\,. \end{aligned} \tag{111}$$

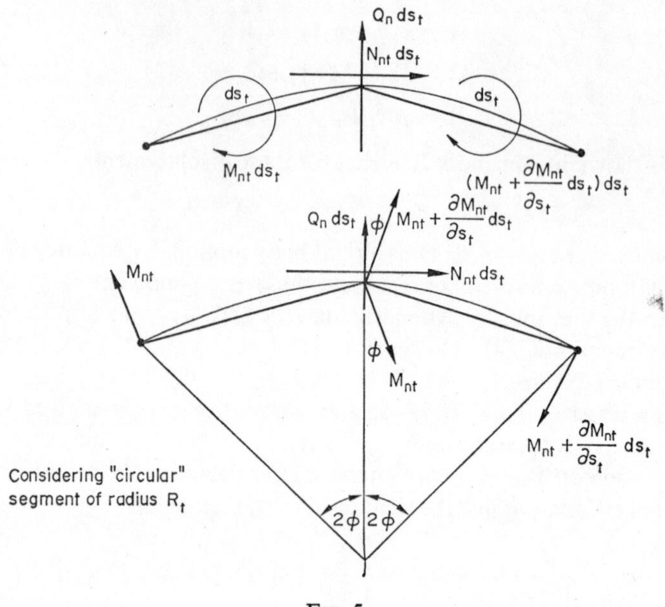

FIG. 5.

In Fig. 5 we show a couple of "line elements" of the reference surface at α_n = constant, to show that V_n and T_{nt} are statically equivalent to Q_n, N_{nt}, M_{nt}. Also, note that $A_t \partial\alpha_t \rightarrow \partial s_t$. These line (chord) elements are of the same length as the shell they subtend, and note that $2\phi = ds_t/R_t$.

First we "divide" $M_{nt} ds_t$ and $(M_{nt} + (\partial M_{nt}/\partial s_t) ds_t) ds_t$ into equivalent couples, and then at the center point, consider the projections of the couple components on the normal and tangent to the curve.[5] Then there is an additional normal (vertical) "force" at the center of magnitude

$$\left(M_{nt} + \frac{\partial M_{nt}}{\partial s_t} ds_t\right) \cos\phi - (M_{nt}) \cos\phi$$

but since $\phi \ll 1$, $\cos\phi \cong 1$, so that we have a net force of $(\partial M_{nt}/\partial s_t)\, ds_t$

[5] It is not explicitly stated, but the directions of the forces are really determined by being normal to the chords—thus like angle ϕ with vertical.

acting, added to the transverse shearing force $Q_n ds_t$, giving the *Kirchoff effective shear force*

$$V_n = Q_n + \frac{\partial M_{nt}}{\partial s_t}.$$

If we also project the two center "forces" onto the tangent at the center we get a net result of

$$\left(M_{nt} + \frac{\partial M_{nt}}{\partial s_t} ds_t\right) \sin\phi + M_{nt} \sin\phi \cong 2M_{nt}\phi$$

where we have taken $\phi \cong \sin\phi$, and disregarded the change in M_{nt} compared to M_{nt} itself. Since $ds_t = 2\phi R_t$ we thus have a net horizontal force $M_{nt} ds_t/R_t$, leading to an effective ("in-plane") shear force

$$T_{nt} = N_{nt} + M_{nt}/R_t.$$

This provides a physically (intuitive) appealing explanation for the boundary conditions (111). Obviously, to "rearrange" the twisting moments we are invoking St. Venant's principle, assuming that the "disturbance" thus created will not propagate very deeply into the shell from the boundary curve.

We note four possible support conditions at an edge α_n = constant:

$$\begin{aligned}
&\text{(a) Free edge: } N_n = T_{nt} = V_n = M_n = 0 \\
&\text{(b) Clamped edge: } u_n = u_t = w = \beta_n = 0 \\
&\text{(c) Simply supported edge: } u_n = u_t = w = M_n = 0 \\
&\text{(d) Freely supported edge: } u_n = u_t = V_n = M_n = 0.
\end{aligned} \tag{112}$$

To close this section, we list the principal equations and their numbers:

Equilibrium: Equations (82)

Boundary conditions: Equations (83), (111), (112)

Stress–displacement: Equations (103)

Strain–displacement: Equations (66), (104), 105).

APPENDIX IIIA

Verification of equation (56)

$$(ds)^2 = d\mathbf{r}\cdot d\mathbf{r}+2zd\mathbf{r}\cdot d\mathbf{n}+2dz(d\mathbf{r}\cdot\mathbf{n}) + z^2(d\mathbf{n}\cdot d\mathbf{n})+2zdz(\mathbf{n}\cdot d\mathbf{n})+(dz)^2(\mathbf{n}\cdot\mathbf{n})\,.$$

By equations (7), (14), (20), (23) we can find that

$$
\begin{aligned}
d\mathbf{r}\cdot d\mathbf{r} &= A_1^2(d\alpha_1)^2+A_2^2(d\alpha_2)^2\\
d\mathbf{r}\cdot d\mathbf{n} &= L(d\alpha_1)^2+N(d\alpha_2)^2 = A_1^2/R_1(d\alpha_1)^2+A_2^2/R_2(d\alpha_2)^2\\
d\mathbf{r}\cdot\mathbf{n} &= 0 \text{ (}d\mathbf{r}\text{ tangent to surface, }\mathbf{n}\text{ normal)}\\
d\mathbf{n}\cdot d\mathbf{n} &= (A_1/R_1)^2(d\alpha_1)^2+(A_2/R_2)^2(d\alpha_2)^2\\
d\mathbf{n}\cdot\mathbf{n} &= 0 \text{ (}d\mathbf{n}\text{ tangent to surface, }\mathbf{n}\text{ normal)}\\
\mathbf{n}\cdot\mathbf{n} &= 1\,.
\end{aligned}
$$

Thus it follows that

$$
\begin{aligned}
(ds)^2 &= A_1^2(d\alpha_1)^2+A_2^2(d\alpha_2)^2+2z(A_1^2/R_1)(d\alpha_1)^2\\
&\quad +2z(A_2^2/R_2)(d\alpha_2)^2+z^2(A_1/R_1)^2(d\alpha_1)^2+z^2(A_2/R_2)^2(d\alpha_2)^2\\
&\quad +(dz)^2\\
&= A_1^2(1+z/R_1)^2(d\alpha_1)^2+A_2^2(1+z/R_2)^2(d\alpha_2)^2+(dz)^2\,.
\end{aligned}
$$

APPENDIX IIIB

Strain displacement relations in curvilinear coordinates

To define strain in a set of triply orthogonal curvilinear coordinates $\alpha_1, \alpha_2, \alpha_3$, we begin by defining a position vector $\mathbf{R}(\alpha_1, \alpha_2, \alpha_3)$ such that the length of a line element P_0P in the unstrained body is given by

$$(ds)^2 = (d\mathbf{R}\cdot d\mathbf{R}) = \left(\frac{\partial \mathbf{R}}{\partial \alpha_i} d\alpha_i\right)\cdot\left(\frac{\partial \mathbf{R}}{\partial \alpha_j} d\alpha_j\right)$$

$$= \sum_{i=1}^{3}\left(\frac{\partial \mathbf{R}}{\partial \alpha_i}\cdot\frac{\partial \mathbf{R}}{\partial \alpha_i}\right)(d\alpha_i)^2 \overset{D}{\equiv} \sum_{i=1}^{3} A_i^2(d\alpha_i)^2 . \tag{a}$$

The A_k are the Lamé parameters, and $A_k = A_k(\alpha_1, \alpha_2, \alpha_3)$. Since the coordinates are orthogonal, $(\partial R/\partial \alpha_i)\cdot(\partial R/\partial \alpha_j) = 0$, $i \neq j$, and since

$$A_i^2 = \frac{\partial \mathbf{R}}{\partial \alpha_i}\cdot\frac{\partial \mathbf{R}}{\partial \alpha_i} \equiv \mathbf{R}_{,i}\cdot\mathbf{R}_{,i} \tag{b}$$

then $A_i = |\mathbf{R}_{,i}|$ = magnitude of the vector $\mathbf{R}_{,i}$. Then unit vectors tangent to the α_i are given by $(1/A_i)\mathbf{R}_{,i}$ (no sum). If $\mathbf{u} = (\mathbf{u}_1, \mathbf{u}_2, \mathbf{u}_3)$ is the displacement (of point P_0, say) vector projection on the unit tangent vectors, then the displacement at point P_0 is given as

$$\mathbf{u} = \frac{u_1}{A_1}\mathbf{R}_{,1} + \frac{u_2}{A_2}\mathbf{R}_{,2} + \frac{u_3}{A_3}\mathbf{R}_{,3} = \sum_{k=1}^{3}\frac{u_k}{A_k}\mathbf{R}_{,k} . \tag{c}$$

Now, points P_0 and P in the unstrained body, with coordinates α_i and $\alpha_i + d\alpha_i$, go to points P_0', P' after deformation, with coordinates $\alpha_i + u_i/A_i$, and $\alpha_i + u_i/A_i + d\alpha_i + d(u_i/A_i)$. Then, if $\xi_i \equiv u_i/A_i$, for convenience, we may tabulate the *coordinates* P_0, P, P_0', P' as follows:

$$P_0(\alpha_1, \alpha_2, \alpha_3)$$

$$P(\alpha_1+d\alpha_1, \alpha_2+d\alpha_2, \alpha_3+d\alpha_3)$$

$$P_0'(\alpha_1+\xi_1, \alpha_2+\xi_2, \alpha_3+\xi_3)$$

$$P'(\alpha_1+d\alpha_1+\xi_1+d\xi_1, \alpha_2+d\alpha_2+\xi_2+d\xi_2, \alpha_3+d\alpha_3+\xi_3+d\xi_3)\,. \tag{d}$$

The Lamé parameters at P_0' may be approximated as

$$A_i^2(\alpha_1+\xi_1, \alpha_2+\xi_2, \alpha_3+\xi_3) \cong A_i^2(\alpha_1, \alpha_2, \alpha_3)+\sum_{j=1}^{3}\frac{\partial A_i^2}{\partial \alpha_j}\xi_j\,. \tag{e}$$

Then the length of the deformed line element (ds^*) is given by

$$(ds^*)^2 = \sum_{i=1}^{3} A_i^2(\alpha_1+\xi_1, \alpha_2+\xi_2, \alpha_3+\xi_3)\left(d\alpha_i+\sum_{j=1}^{3}\frac{\partial \xi_i}{\partial \alpha_j}d\alpha_j\right)^2. \tag{f}$$

Substituting the expansions (e) and keeping only the linear terms in ξ_k, we find that

$$(ds^*)^2 = \sum_{i=1}^{3} A_i^2(\alpha_1, \alpha_2, \alpha_3)+\sum_{i=1}^{3}\sum_{j=1}^{3}\left[\frac{\partial A_i^2}{\partial \alpha_j}\xi_j(d\alpha_i)^2 + A_i^2\frac{\partial \xi_i}{\partial \alpha_j}d\alpha_j d\alpha_i + A_j^2\frac{\partial \xi_j}{\partial \alpha_i}d\alpha_i d\alpha_j\right]. \tag{g}$$

Then combining equations (a), (g), we find

$$(ds^*)^2-(ds)^2 = \sum_{i=1}^{3}\sum_{j=1}^{3}\left[\frac{\partial A_i^2}{\partial \alpha_j}\xi_j(d\alpha_i)^2 + \left(A_i^2\frac{\partial \xi_i}{\partial \alpha_j}+A_j^2\frac{\partial \xi_j}{\partial \alpha_i}\right)(d\alpha_i d\alpha_j)\right]. \tag{h}$$

Or

$$(ds^*)^2-(ds)^2 = \sum_{i=1}^{3}\sum_{j=1}^{3}\left[\sum_{k=1}^{3}\frac{\partial A_i^2}{\partial \alpha_k}\xi_k(d\alpha_i d\alpha_j)\delta_{ij} + \left(A_i^2\frac{\partial \xi_i}{\partial \alpha_j}+A_j^2\frac{\partial \xi_j}{\partial \alpha_i}\right)(d\alpha_i d\alpha_j)\right]$$

$$= \sum_{i=1}^{3} \sum_{j=1}^{3} \left[\sum_{k=1}^{3} \frac{\partial A_i^2}{\partial \alpha_k} \xi_k \delta_{ij} + A_i^2 \frac{\partial \xi_i}{\partial \alpha_j} + A_j^2 \frac{\partial \xi_j}{\partial \alpha_i} \right] (d\alpha_i d\alpha_j)$$

or

$$\frac{(ds^*)^2 - (ds)^2}{(ds)^2} = \sum_{i=1}^{3} \sum_{j=1}^{3} \left[\sum_{k=1}^{3} \frac{\partial A_i^2}{\partial \alpha_k} \xi_k \delta_{ij} + A_i^2 \frac{\partial \xi_i}{\partial \alpha_j} + A_j^2 \frac{\partial \xi_j}{\partial \alpha_i} \right] \left(\frac{d\alpha_i}{ds} \right) \left(\frac{d\alpha_j}{ds} \right). \tag{i}$$

Now we note the following definitions. *The extension ε of a line element originally of length ds, now of length ds* after deformation, is*

$$\varepsilon = \frac{ds^* - ds}{ds}. \tag{j}$$

Further, the *strain tensor* ε_{ij} is defined by

$$(ds^*)^2 - (ds)^2 = \sum_{i=1}^{3} \sum_{j=1}^{3} 2\varepsilon_{ij} l_i l_j (ds)^2 \tag{k}$$

where l_i are the direction cosines of the element (ds) with respect to the $(d\alpha_i)$, here given by

$$l_k = A_k \frac{d\alpha_k}{ds}, \quad \text{no sum on } k. \tag{l}$$

Note that

$$\sum_{k=1}^{3} l_k^2 = \sum_{k=1}^{3} A_k^2 \left(\frac{d\alpha_k}{ds} \right)^2 = 1$$

by virtue of equation (a), and that $l_i l_k = 0$ by virtue of the orthogonality of the curvilinear coordinates.

Now from equation (j), $ds^*/ds = 1 + \varepsilon$, so that the left side of equations (k) or (i) can be written as

$$(ds^*)^2 - (ds)^2 = [(1+\varepsilon)^2 - 1](ds)^2 = (\varepsilon^2 + 2\varepsilon)(ds)^2. \tag{m}$$

To a first approximation, for $\varepsilon \ll 1$,

$$(ds^*)^2 - (ds)^2 = 2\varepsilon(ds)^2. \tag{n}$$

So that combining equations (n), (k), (i) we find that

$$2\varepsilon(ds)^2 = \sum_{i=1}^{3} \sum_{j=1}^{3} 2\varepsilon_{ij} l_i l_j (ds)^2$$

$$= \sum_{i=1}^{3} \sum_{j=1}^{3} \left[\sum_{k=1}^{3} \frac{\partial A_i^2}{\partial \alpha_k} \xi_k \delta_{ij} + A_i^2 \frac{\partial \xi_i}{\partial \alpha_j} + A_j^2 \frac{\partial \xi_j}{\partial \alpha_i} \right] (d\alpha_i d\alpha_j)$$

so that the components ε_{ij} of the strain tensor are defined *from*

$$\varepsilon = \sum_{i=1}^{3} \sum_{j=1}^{3} \varepsilon_{ij} l_i l_j \tag{o}$$

and, using equations (1), are given by

$$2\varepsilon_{ij} = \frac{1}{A_i A_j} \left[\sum_{k=1}^{3} \frac{\partial A_i^2}{\partial \alpha_k} \xi_k \delta_{ij} + A_i^2 \frac{\partial \xi_i}{\partial \alpha_j} + A_j^2 \frac{\partial \xi_j}{\partial \alpha_i} \right]. \tag{p}$$

For *normal strains*, ε_{ii}, $i = j$, then

$$\varepsilon_{ii} = \frac{\partial}{\partial \alpha_i} \left(\frac{U_i}{A_i} \right) + \frac{1}{2A_i^2} \sum_{k=1}^{3} \frac{\partial A_i}{\partial \alpha_k} \left(\frac{U_k}{A_k} \right). \tag{q}$$

For *shearing strains*, $\gamma_{ij} = 2\varepsilon_{ij}$, $i \neq j$, then

$$\gamma_{ij} = \frac{A_i}{A_j} \frac{\partial}{\partial \alpha_j} \left(\frac{U_i}{A_i} \right) + \frac{A_j}{A_i} \frac{\partial}{\partial \alpha_i} \left(\frac{U_j}{A_j} \right). \tag{r}$$

For *the shell curvilinear coordinate system*, we identify the following

$\alpha_1 = \alpha_1$	$\alpha_2 = \alpha_2$	$\alpha_3 = z$
$A_1 = A_1(1 + a/R_1)$	$A_2 = A_2(1 + z/R_2)$	$A_3 = 1$
$U_1 = U_1$	$U_2 = U_2$	$U_3 = W$.

APPENDIX IIIC

Verification of equation (67)

START with equations (65):

$$\frac{\partial w}{\partial \alpha_1} = \frac{A_1}{R_1} u_1 - A_1\beta_1; \quad \frac{\partial w}{\partial \alpha_2} = \frac{A_2}{R_2} u_2 - A_2\beta_2 .$$

Thus

$$\frac{\partial}{\partial \alpha_2}\left(\frac{A_1}{R_1} u_1 - A_1\beta_1\right) - \frac{\partial}{\partial \alpha_1}\left(\frac{A_2}{R_2} u_2 - A_2\beta_2\right) = 0$$

or

$$\frac{A_1}{R_1}\frac{\partial u_1}{\partial \alpha_2} - \frac{A_2}{R_2}\frac{\partial u_2}{\partial \alpha_1} + u_1 \frac{\partial}{\partial \alpha_2}\left(\frac{A_1}{R_1}\right) - u_2 \frac{\partial}{\partial \alpha_1}\left(\frac{A_2}{R_2}\right)$$

$$= \frac{\partial}{\partial \alpha_2}(A_1\beta_1) - \frac{\partial}{\partial \alpha_1}(A_2\beta_2).$$

After using the Codazzi conditions (28):

$$\frac{A_1}{R_1}\frac{\partial u_1}{\partial \alpha_2} - \frac{u_2}{R_1}\frac{\partial A_2}{\partial \alpha_1} - \frac{A_2}{R_2}\frac{\partial u_1}{\partial \alpha_1} + \frac{u_1}{R_2}\frac{\partial A_1}{\partial \alpha_2}$$

$$= A_1 \frac{\partial \beta_1}{\partial \alpha_2} - \beta_2 \frac{\partial A_2}{\partial \alpha_1} - A_2 \frac{\partial \beta_2}{\partial \alpha_1} + \beta_1 \frac{\partial A_1}{\partial \alpha_2}$$

or, after a trivial manipulation,

$$\frac{A_1A_2}{R_1}\left[\frac{1}{A_2}\frac{\partial u_1}{\partial \alpha_2} - \frac{u_2}{A_1A_2}\frac{\partial A_2}{\partial \alpha_1}\right] - \frac{A_1A_2}{R_2}\left[\frac{1}{A_1}\frac{\partial u_2}{\partial \alpha_1} - \frac{u_1}{A_2A_2}\frac{\partial A_1}{\partial \alpha_2}\right]$$

$$= A_1A_2\left[\frac{1}{A_2}\frac{\partial \beta_1}{\partial \alpha_2} - \frac{\beta_2}{A_1A_2}\frac{\partial A_2}{\partial \alpha_1} - \left(\frac{1}{A_1}\frac{\partial \beta_2}{\partial \alpha_1} - \frac{\beta_1}{A_1A_2}\frac{\partial A_1}{\partial \alpha_2}\right)\right].$$

Then by equations (66i, j, k, l):

$$\frac{\omega_2}{R_1} - \frac{\omega_1}{R_2} = \tau_2 - \tau_1$$

or

$$\tau_2 + \frac{\omega_1}{R_2} = \tau_1 + \frac{\omega_2}{R_1}.$$

APPENDIX IIID

Alternate derivation of the equilibrium equations

THIS addendum is to carry out the first variational procedure defined on p. 34, i.e., to perform the variation indicated, say, in equation (81), with $Q_1 \to 0$, $Q_2 \to 0$, and the β_1, β_2 terms replaced from equations (65); then there will only be three independent variations, i.e., δu_1, δu_2 and δw. Then

$$\delta U_e = \iint \Big[N_1 \Big\{ A_2 \frac{\partial \delta u_1}{\partial \alpha_1} + \frac{\partial A_1}{\partial \alpha_2} \delta u_2 + A_1 A_2 \frac{\delta w}{R_1} \Big\}$$

$$+ N_2 \Big\{ A_1 \frac{\partial \delta u_2}{\partial \alpha_2} + \frac{\partial A_2}{\partial \alpha_1} \delta u_1 + A_1 A_2 \frac{\delta w}{R_2} \Big\} + N_{12} \Big\{ A_2 \frac{\partial \delta u_2}{\partial \alpha_1}$$

$$- \frac{\partial A_1}{\partial \alpha_2} \delta u_1 \Big\} + N_{21} \Big\{ A_1 \frac{\partial \delta u_1}{\partial \alpha_2} - \frac{\partial A_2}{\partial \alpha_1} \delta u_2 \Big\} + M_1 \Big\{ \frac{\partial A_1}{\partial \alpha_2} \Big(\frac{\delta u_2}{R_2}$$

$$- \frac{1}{A_2} \frac{\partial \delta w}{\partial \alpha_2} \Big) + A_2 \frac{\partial}{\partial \alpha_1} \Big(\frac{\delta u_1}{R_1} - \frac{1}{A_1} \frac{\partial \delta w}{\partial \alpha_1} \Big) \Big\} + M_2 \Big\{ \frac{\partial A_2}{\partial \alpha_1} \Big(\frac{\delta u_1}{R_1}$$

$$- \frac{1}{A_1} \frac{\partial \delta w}{\partial \alpha_1} \Big) + A_1 \frac{\partial}{\partial \alpha_2} \Big(\frac{\delta u_2}{R_2} - \frac{1}{A_2} \frac{\partial \delta w}{\partial \alpha_2} \Big) \Big\} + M_{12} \Big\{ - \frac{\partial A_1}{\partial \alpha_2} \Big(\frac{\delta u_1}{R_1}$$

$$- \frac{1}{A_1} \frac{\partial \delta w}{\partial \alpha_1} \Big) + A_2 \frac{\partial}{\partial \alpha_1} \Big(\frac{\delta u_2}{R_2} - \frac{1}{A_2} \frac{\partial \delta w}{\partial \alpha_2} \Big) \Big\} + M_{21} \Big\{ - \frac{\partial A_2}{\partial \alpha_1} \Big(\frac{\delta u_2}{R_2}$$

$$- \frac{1}{A_2} \frac{\partial \delta w}{\partial \alpha_2} \Big) + A_1 \frac{\partial}{\partial \alpha_2} \Big(\frac{\delta u_1}{R_1} - \frac{1}{A_1} \frac{\partial \delta w}{\partial \alpha_1} \Big) \Big\} \Big] d\alpha_1 d\alpha_2 = 0 .$$

In what follows we shall ignore the boundary conditions. Then

$$\delta U_e = \iint\Big[-\frac{\partial}{\partial\alpha_1}(N_1A_2)\delta u_1 + N_1\frac{\partial A_1}{\partial\alpha_2}\delta u_2 + \frac{N_1}{R_1}A_1A_2\delta w$$

$$-\frac{\partial}{\partial\alpha_2}(N_2A_1)\delta u_2 + N_2\frac{\partial A_2}{\partial\alpha_1}\delta u_1 + \frac{N_2}{R_2}A_1A_2\delta w -$$

$$-\frac{\partial}{\partial\alpha_1}(N_{12}A_2)\delta u_2 - N_{12}\frac{\partial A_1}{\partial\alpha_2}\delta u_1 - \frac{\partial}{\partial\alpha_2}(N_{21}A_1)\delta u_1$$

$$-N_{21}\frac{\partial A_2}{\partial\alpha_1}\delta u_2 + \frac{M_1}{R_2}\frac{\partial A_1}{\partial\alpha_2}\delta u_2 + \frac{\partial}{\partial\alpha_2}\left(\frac{M_1}{A_2}\frac{\partial A_1}{\partial\alpha_2}\right)\delta w$$

$$-\frac{\partial}{\partial\alpha_1}(M_1A_2)\left(\frac{\delta u_1}{R_1} - \frac{1}{A_1}\frac{\partial\delta w}{\partial\alpha_1}\right) + \frac{M_2}{R_1}\frac{\partial A_2}{\partial\alpha_1}\delta u_1$$

$$+\frac{\partial}{\partial\alpha_1}\left(\frac{M_2}{A_1}\frac{\partial A_2}{\partial\alpha_1}\right)\delta w - \frac{\partial}{\partial\alpha_2}(M_2A_1)\left(\frac{\delta u_2}{R_2} - \frac{1}{A_2}\frac{\partial\delta w}{\partial\alpha_2}\right)$$

$$-\frac{M_{12}}{R_1}\frac{\partial A_1}{\partial\alpha_2}\delta u_1 - \frac{\partial}{\partial\alpha_1}\left(\frac{M_{12}}{A_1}\frac{\partial A_1}{\partial\alpha_2}\right)\delta w$$

$$-\frac{\partial}{\partial\alpha_1}(M_{12}A_2)\left(\frac{\delta u_2}{R_2} - \frac{1}{A_2}\frac{\partial\delta w}{\partial\alpha_2}\right) - \frac{M_{21}}{R_2}\frac{\partial A_2}{\partial\alpha_1}\delta u_2$$

$$-\frac{\partial}{\partial\alpha_2}\left(\frac{M_{21}}{A_2}\frac{\partial A_2}{\partial\alpha_1}\right)\delta w - \frac{\partial}{\partial\alpha_2}(M_{21}A_1)\left(\frac{\delta u_1}{R_1} - \frac{1}{A_1}\frac{\partial\delta w}{\partial\alpha_1}\right)\Big] \times$$

$$\times\, d\alpha_1 d\alpha_2 = 0\,.$$

Or, regrouping some terms:

$$\delta U_e = \iint\Big\{\Big[-\frac{\partial}{\partial\alpha_1}(N_1A_2) + N_2\frac{\partial A_2}{\partial\alpha_1} - N_{12}\frac{\partial A_1}{\partial\alpha_2} - \frac{\partial}{\partial\alpha_2}(N_{21}A_1)$$

$$-\frac{1}{R_1}\frac{\partial}{\partial\alpha_1}(M_1A_2) + \frac{M_2}{R_1}\frac{\partial A_2}{\partial\alpha_1} - \frac{M_{12}}{R_1}\frac{\partial A_1}{\partial\alpha_2} - \frac{1}{R_1}\frac{\partial}{\partial\alpha_2}(M_{21}A_1)\Big]\delta u_1$$

$$+\Big[N_1\frac{\partial A_1}{\partial\alpha_2} - \frac{\partial}{\partial\alpha_2}(N_2A_1) - \frac{\partial}{\partial\alpha_1}(N_{12}A_2) - N_{21}\frac{\partial A_2}{\partial\alpha_1}$$

$$
+\frac{M_1}{R_2}\frac{\partial A_1}{\partial \alpha_2}-\frac{1}{R_2}\frac{\partial}{\partial \alpha_2}(M_2A_1)-\frac{1}{R_2}\frac{\partial}{\partial \alpha_1}(M_{12}A_2)
$$

$$
-\frac{M_{21}}{R_2}\frac{\partial A_2}{\partial \alpha_1}\Bigg]\delta u_2+\Bigg[\left(\frac{N_1}{R_1}+\frac{N_2}{R_2}\right)A_1A_2+\frac{\partial}{\partial \alpha_2}\left(\frac{M_1}{A_2}\frac{\partial A_1}{\partial \alpha_2}\right)
$$

$$
-\frac{\partial}{\partial \alpha_1}\left(\frac{1}{A_1}\frac{\partial}{\partial \alpha_1}(M_1A_2)\right)+\frac{\partial}{\partial \alpha_1}\left(\frac{M_2}{A_1}\frac{\partial A_2}{\partial \alpha_1}\right)
$$

$$
-\frac{\partial}{\partial \alpha_2}\left(\frac{1}{A_2}\frac{\partial}{\partial \alpha_2}(M_2A_1)\right)-\frac{\partial}{\partial \alpha_1}\left(\frac{M_{12}}{A_1}\frac{\partial A_1}{\partial \alpha_2}\right)
$$

$$
-\frac{\partial}{\partial \alpha_2}\left(\frac{1}{A_2}\frac{\partial}{\partial \alpha_1}(M_{12}A_2)\right)-\frac{\partial}{\partial \alpha_2}\left(\frac{M_{21}}{A_2}\frac{\partial A_2}{\partial \alpha_1}\right)
$$

$$
-\frac{\partial}{\partial \alpha_1}\left(\frac{1}{A_1}\frac{\partial}{\partial \alpha_2}(M_{21}A_1)\right)\Bigg]\delta w\Bigg\}d\alpha_1 d\alpha_2;
$$

$$
-\delta U_e = 0 = \iint\Bigg\{\Bigg[\frac{\partial}{\partial \alpha_1}(N_1A_2)+\frac{\partial}{\partial \alpha_2}(N_{21}A_1)+N_{12}\frac{\partial A_1}{\partial \alpha_2}
$$

$$
-N_2\frac{\partial A_2}{\partial \alpha_1}+\frac{1}{R_1}\frac{\partial}{\partial \alpha_1}(M_1A_2)+\frac{1}{R_1}\frac{\partial}{\partial \alpha_2}(M_{21}A_1)
$$

$$
+\frac{M_{12}}{R_1}\frac{\partial A_1}{\partial \alpha_2}-\frac{M_2}{R_1}\frac{\partial A_2}{\partial \alpha_1}\Bigg]\delta u_1+\Bigg[\frac{\partial}{\partial \alpha_1}(N_{12}A_2)
$$

$$
+\frac{\partial}{\partial \alpha_2}(N_2A_1)+N_{21}\frac{\partial A_2}{\partial \alpha_1}-N_1\frac{\partial A_1}{\partial \alpha_2}+\frac{1}{R_2}\frac{\partial}{\partial \alpha_1}(M_{12}A_2)
$$

$$
+\frac{1}{R_2}\frac{\partial}{\partial \alpha_2}(M_2A_1)+\frac{M_{21}}{R_2}\frac{\partial A_2}{\partial \alpha_1}-\frac{M_1}{R_2}\frac{\partial A_1}{\partial \alpha_2}\Bigg]\delta u_2
$$

$$
+\Bigg[\frac{\partial}{\partial \alpha_1}\frac{1}{A_1}\left(\frac{\partial}{\partial \alpha_1}(M_1A_2)+\frac{\partial}{\partial \alpha_2}(M_{21}A_1)+M_{12}\frac{\partial A_1}{\partial \alpha_2}-M^2\frac{\partial A_2}{\partial \alpha_1}\right)
$$

$$
+\frac{\partial}{\partial \alpha_2}\frac{1}{A_2}\left(\frac{\partial}{\partial \alpha_1}(M_{12}A_2)+\frac{\partial}{\partial \alpha_2}(M_2A_1)+M_{21}\frac{\partial A_2}{\partial \alpha_1}-M_1\frac{\partial A_1}{\partial \alpha_2}\right)
$$

$$
-\left(\frac{N_1}{R_1}+\frac{N_2}{R_2}\right)A_1A_2\Bigg]\delta w\Bigg\}d\alpha_1 d\alpha_2 .
$$

APPENDIX IIIE

Strain parameter values for rigid body motions

WE want to evaluate ε_i^0, κ_i, ω, τ, and τ^* for the rigid body motions defined by equations (108), (109). Thus for

$$\mathbf{U} = \mathbf{\Delta} + \mathbf{\Omega} \times \mathbf{R}$$

since $\mathbf{\Delta}$, $\mathbf{\Omega}$ are constants we have

$$\frac{\partial \mathbf{\Delta}}{\partial \alpha_1} = \frac{\partial \mathbf{\Delta}}{\partial \alpha_2} = \frac{\partial \mathbf{\Omega}}{\partial \alpha_1} = \frac{\partial \mathbf{\Omega}}{\partial \alpha_2} = 0 .$$

Now, for example,

$$\frac{\partial \mathbf{\Delta}}{\partial \alpha_1} = \delta_{1,1}\mathbf{t}_1 + \delta_{2,1}\mathbf{t}_2 + \delta_{n,1}\mathbf{n} + \delta_1 \mathbf{t}_{1,1} + \delta_2 \mathbf{t}_{2,1} + \delta_n \mathbf{n}_{,1} = 0 .$$

By virtue of equations (27), (23) we can write

$$\frac{\partial \mathbf{\Delta}}{\partial \alpha_1} = \left(\delta_{1,1} + \frac{\delta_2}{A_2}\frac{\partial A_1}{\partial \alpha_2} + \delta_n \frac{A_1}{R_1}\right)\mathbf{t}_1 + \left(\delta_{2,1} - \frac{\delta_1}{A_2}\frac{\partial A_1}{\partial \alpha_2}\right)\mathbf{t}_2 + \left(\delta_{n,1} - \delta_1 \frac{A_1}{R_1}\right)\mathbf{n} = 0$$

which implies that

$$\delta_{1,1} = -\frac{\delta_2}{A_2}\frac{\partial A_1}{\partial \alpha_2} - \delta_n \frac{A_1}{R_1}, \quad \delta_{2,1} = \frac{\delta_1}{A_2}\frac{\partial A_1}{\partial \alpha_2}$$

$$\delta_{n,1} = \delta_1 \frac{A_1}{R_1} . \tag{a}$$

Similarly, from $\partial\mathbf{\Delta}/\partial\alpha_2 = 0$ we would obtain,

$$\delta_{1,2} = \frac{\delta_2}{A_1}\frac{\partial A_2}{\partial\alpha_1}, \quad \delta_{2,2} = -\frac{\delta_1}{A_1}\frac{\partial A_2}{\partial\alpha_1} - \delta_n\frac{A_2}{R_2} \tag{b}$$

$$\delta_{n,2} = \delta_2\frac{A_2}{R_2}.$$

The consequences of $\mathbf{\Omega}$ being a constant vector are immediately obtained from the results (a, b) by replacing $\delta_1 \to -\Omega_2$, $\delta_2 \to \Omega_1$, $\delta_n \to \Omega_n$ [see equations (109)]. Then

$$\Omega_{1,1} = -\frac{\Omega_2}{A_2}\frac{\partial A_1}{\partial\alpha_2}$$

$$\Omega_{1,2} = \frac{\Omega_2}{A_1}\frac{\partial A_2}{\partial\alpha_1} - \Omega_n\frac{A_2}{R_2}$$

$$\Omega_{2,1} = \frac{\Omega_1}{A_2}\frac{\partial A_1}{\partial\alpha_2} + \Omega_n\frac{A_1}{R_1} \tag{c}$$

$$\Omega_{2,2} = -\frac{\Omega_1}{A_1}\frac{\partial A_2}{\partial\alpha_1}$$

$$\Omega_{n,1} = -\Omega_2\frac{A_1}{R_1}$$

$$\Omega_{n,2} = \Omega_1\frac{A_2}{R_2}.$$

Now the vector $\mathbf{R}$ is the position vector to a point on the middle surface, hence

$$\mathbf{R} = \mathbf{r}(\alpha_1, \alpha_2) = \rho_1\mathbf{t}_1 + \rho_2\mathbf{t}_2 + \rho_n\mathbf{n} \tag{d}$$

From equations (21) we have

$$\frac{\partial\mathbf{r}}{\partial\alpha_1} = A_1\mathbf{t}_1, \quad \frac{\partial\mathbf{r}}{\partial\alpha_2} = A_2\mathbf{t}_2 \tag{e}$$

so that by differentiating equation (d)

$$\frac{\partial\mathbf{r}}{\partial\alpha_1} \equiv A_1\mathbf{t}_1 = \rho_{1,1}\mathbf{t}_1 + \rho_1\mathbf{t}_{1,1} + \rho_{2,1}\mathbf{t}_2 + \rho_{n,1}\mathbf{n} + \rho_2\mathbf{t}_{2,1} + \rho_n\mathbf{n}_{,1}.$$

Therefore

$$A_1\mathbf{t}_1 = \left(\rho_{1,1}+\frac{\rho_2}{A_2}\frac{\partial A_1}{\partial \alpha_2}+\rho_n\frac{A_1}{R_1}\right)\mathbf{t}_1$$

$$+\left(\rho_{2,1}-\frac{\rho_1}{A_2}\frac{\partial A_1}{\partial \alpha_2}\right)\mathbf{t}_2+\left(\rho_{n,1}-\rho_1\frac{A_1}{R_1}\right)\mathbf{n} \qquad \text{(f)}$$

Equating vector components yields (and including the analogous computation for $\partial\mathbf{r}/\partial\alpha_2$)

$$\rho_{1,1} = A_1-\frac{\rho_2}{A_2}\frac{\partial A_1}{\partial \alpha_2}-\rho_n\frac{A_1}{R_1}$$

$$\rho_{2,1} = \frac{\rho_1}{A_2}\frac{\partial A_1}{\partial \alpha_2}, \quad \rho_{n,1} = \rho_1\frac{A_1}{R_1}$$

$$\rho_{1,2} = \frac{\rho_2}{A_1}\frac{\partial A_2}{\partial \alpha_1}, \quad \rho_{n,2} = \rho_2\frac{A_2}{R_2} \qquad \text{(g)}$$

$$\rho_{2,2} = A_2-\frac{\rho_1}{A_1}\frac{\partial A_2}{\partial \alpha_1}-\rho_n\frac{A_2}{R_2}.$$

Finally, from

$$\mathbf{U} = \boldsymbol{\Delta}+\boldsymbol{\Omega}\times\mathbf{R} = \boldsymbol{\Delta}+\boldsymbol{\Omega}\times\mathbf{r}$$

we have

$$\mathbf{U} = (\delta_1+\Omega_1\rho_n-\Omega_n\rho_2)\mathbf{t}_1+(\delta_2+\Omega_n\rho_1+\Omega_2\rho_n)\mathbf{t}_2$$

$$+(\delta_n-\Omega_2\rho_2-\Omega_1\rho_1)\mathbf{n}$$

so that by comparison with equations (64) with $z = 0$, for the surface,

$$u_1 = \delta_1+\Omega_1\rho_n-\Omega_n\rho_2$$

$$u_2 = \delta_2+\Omega_n\rho_1+\Omega_2\rho_n \qquad \text{(h)}$$

$$w = \delta_n-\Omega_2\rho_2-\Omega_1\rho_1.$$

Then from equations (66)

$$\varepsilon_1^0 = \frac{1}{A_1}\frac{\partial u_1}{\partial \alpha_1}+\frac{u_2}{A_1A_2}\frac{\partial A_2}{\partial \alpha_2}+\frac{w}{R_1}$$

$$= \frac{1}{A_1}(\delta_{1,1}+\Omega_{1,1}\rho_n+\Omega_1\rho_{n,1}-\Omega_{n\ 1}\rho_2$$

$$-\Omega_n\rho_{2,1})+\frac{1}{A_1A_2}\frac{\partial A_1}{\partial\alpha_2}(\delta_2+\Omega_n\rho_1+\Omega_2\rho_n)$$

$$+\frac{1}{R_1}(\delta_n-\Omega_2\rho_2-\Omega_1\rho_1)$$

or using equations (a), (c), (g), we have

$$\varepsilon_1^0 = -\frac{\delta_2}{A_1A_2}\frac{\partial A_1}{\partial\alpha_2}-\frac{\delta_n}{R_1}-\frac{\rho_n\Omega_2}{A_1A_2}\frac{\partial A_1}{\partial\alpha_2}+\frac{\Omega_1\rho_1}{R_1}$$

$$+\frac{\Omega_2\rho_2}{R_1}-\frac{\Omega_n\rho_1}{A_1A_2}\frac{\partial A_1}{\partial\alpha_2}+\frac{\delta_2}{A_1A_2}\frac{\partial A_1}{\partial\alpha_2}$$

$$+\frac{\Omega_n\rho_1}{A_1A_2}\frac{\partial A_1}{\partial\alpha_2}+\frac{\Omega_2\rho_n}{A_1A_2}\frac{\partial A_1}{\partial\alpha_2}+\frac{\delta_n}{R_1}-\frac{\Omega_2\rho_2}{R_1}-\frac{\Omega_1\rho_1}{R_1}.$$

Therefore

$$\varepsilon_1^0 = 0.$$

Similarly, $\varepsilon_2^0 = 0$. Further

$$\omega_1 = \frac{1}{A_1}\frac{\partial u_2}{\partial\alpha_1}-\frac{u_1}{A_1A_2}\frac{\partial A_1}{\partial\alpha_2}$$

$$= \frac{1}{A_1}(\delta_{2,1}+\rho_1\Omega_{n,1}+\Omega_n\rho_{1,1}+\rho_n\Omega_{2,1}+\Omega_2\rho_{n,1})$$

$$-\frac{1}{A_1A_2}\frac{\partial A_1}{\partial\alpha_2}(\delta_1+\Omega_1\rho_n-\Omega_n\rho_2)$$

$$= \frac{\delta_1}{A_1A_2}\frac{\partial A_1}{\partial\alpha_2}-\frac{\rho_1\Omega_2}{R_1}+\Omega_n-\frac{\rho_2\Omega_n}{A_1A_2}\frac{\partial A_1}{\partial\alpha_2}$$

$$-\frac{\rho_n\Omega_n}{R_1}+\frac{\rho_n\Omega_1}{A_1A_2}\frac{\partial A_1}{\partial\alpha_2}+\frac{\rho_n\Omega_n}{R_1}+\frac{\Omega_2\rho_1}{R_1}$$

$$-\frac{\delta_1}{A_1A_2}\frac{\partial A_1}{\partial\alpha_2}-\frac{\Omega_1\rho_n}{A_1A_2}\frac{\partial A_1}{\partial\alpha_2}+\frac{\Omega_n\rho_2}{A_1A_2}\frac{\partial A_1}{\partial\alpha_2}$$

$$= \Omega_n$$

and also

$$\omega_2 = \frac{1}{A_2}\frac{\partial u_1}{\partial \alpha_2} - \frac{u_2}{A_1 A_2}\frac{\partial A_2}{\partial \alpha_1}$$

$$= \frac{1}{A_2}(\delta_{1,2} - \Omega_n \rho_{2,2} - \rho_2 \Omega_{n,2} + \Omega_1 \rho_{n,2} + \rho_n \Omega_{1,2})$$

$$- \frac{1}{A_1 A_2}\frac{\partial A_2}{\partial \alpha_1}(\delta_2 + \Omega_n \rho_1 + \Omega_2 \rho_n)$$

$$= \frac{\delta_2}{A_1 A_2}\frac{\partial A_2}{\partial \alpha_1} - \Omega_n + \frac{\rho_1 \Omega_n}{A_1 A_2}\frac{\partial A_2}{\partial \alpha_1} + \frac{\Omega_n \rho_n}{R_2}$$

$$- \frac{\rho_2 \Omega_1}{R_2} + \frac{\Omega_1 \rho_2}{R_2} + \frac{\rho_n \Omega_2}{A_1 A_2}\frac{\partial A_2}{\partial \alpha_1} - \frac{\rho_n \Omega_n}{R_2}$$

$$- \frac{\delta_2}{A_1 A_2}\frac{\partial A_2}{\partial \alpha_1} - \frac{\rho_1 \Omega_n}{A_1 A_2}\frac{\partial A_2}{\partial \alpha_1} - \frac{\rho_n \Omega_2}{A_1 A_2}\frac{\partial A_2}{\partial \alpha_1}$$

$$= -\Omega_n$$

so that $\omega = \omega_1 + \omega_2 = \Omega_n - \Omega_n \equiv 0$.

Now

$$\beta_1 = \frac{u_1}{R_1} - \frac{1}{A_1}\frac{\partial w}{\partial \alpha_1}$$

$$= \frac{\delta_1}{R_1} + \frac{\Omega_1 \rho_n}{R_1} - \frac{\Omega_n \rho_2}{R_1} - \frac{1}{A_1}(\delta_{n,1} - \Omega_2 \rho_{2,1} - \rho_2 \Omega_{2,1} - \Omega_{1,1}\rho_1 - \rho_{1,1}\Omega_1)$$

$$= \frac{\delta_1}{R_1} + \frac{\Omega_1 \rho_n}{R_1} - \frac{\Omega_n \rho_2}{R_1} - \frac{\delta_1}{R_1} + \frac{\Omega_2 \rho_1}{A_1 A_2}\frac{\partial A_1}{\partial \alpha_2} + \frac{\rho_2 \Omega_1}{A_1 A_2}\frac{\partial A_1}{\partial \alpha_2}$$

$$+ \frac{\rho_2 \Omega_n}{R_1} - \frac{\rho_1 \Omega_2}{A_1 A_2}\frac{\partial A_1}{\partial \alpha_2} + \Omega_1 - \frac{\Omega_1 \rho_2}{A_1 A_2}\frac{\partial A_1}{\partial \alpha_2} - \frac{\rho_n \Omega_1}{R_1}.$$

Therefore $\beta_1 = \Omega_1$.

And similarly, $\beta_2 = \Omega_2$.

Then the curvatures are

$$\kappa_1 = \frac{1}{A_1}\frac{\partial \beta_1}{\partial \alpha_1} + \frac{\beta_2}{A_1 A_2}\frac{\partial A_1}{\partial \alpha_2} = \frac{1}{A_1}\frac{\partial \Omega_1}{\partial \alpha_1} + \frac{\Omega_2}{A_1 A_2}\frac{\partial A_1}{\partial \alpha_2}$$

$$= -\frac{\Omega_2}{A_1 A_2}\frac{\partial A_1}{\partial \alpha_2} + \frac{\Omega_2}{A_1 A_2}\frac{\partial A_1}{\partial \alpha_2} = 0\,.$$

$$\kappa_2 = \frac{1}{A_2}\frac{\partial \beta_2}{\partial \alpha_2} + \frac{\beta_1}{A_1 A_2}\frac{\partial A_2}{\partial \alpha_1} = \frac{1}{A_2}\frac{\partial \Omega_2}{\partial \alpha_2} + \frac{\Omega_1}{A_1 A_2}\frac{\partial A_2}{\partial \alpha_1}$$

$$= -\frac{\Omega_1}{A_1 A_2}\frac{\partial A_2}{\partial \alpha_1} + \frac{\Omega_1}{A_1 A_2}\frac{\partial A_2}{\partial \alpha_1} = 0\,.$$

Thus, so far, we have found that

$$\varepsilon_1^0 = \varepsilon_2^0 = \kappa_1 = \kappa_2 = 0, \quad \omega_1 = \Omega_n, \omega_2 = -\Omega_n$$

and

$$\omega = \omega_1 + \omega_2 = 0, \quad \beta_1 = \Omega_1, \beta_2 = \Omega_2. \tag{i}$$

Continuing we find

$$\tau_1 = \frac{1}{A_1}\frac{\partial \beta_2}{\partial \alpha_1} - \frac{\beta_1}{A_1 A_2}\frac{\partial A_1}{\partial \alpha_2} = \frac{1}{A_1}\frac{\partial \Omega_2}{\partial \alpha_1} - \frac{\Omega_1}{A_1 A_2}\frac{\partial A_1}{\partial \alpha_2}$$

$$= \frac{\Omega_1}{A_1 A_2}\frac{\partial A_1}{\partial \alpha_2} + \frac{\Omega_n}{R_1} - \frac{\Omega_1}{A_1 A_2}\frac{\partial A_1}{\partial \alpha_2} = \frac{\Omega_n}{R_1},$$

$$\tau_2 = \frac{1}{A_2}\frac{\partial \beta_1}{\partial \alpha_2} - \frac{\beta_2}{A_1 A_2}\frac{\partial A_2}{\partial \alpha_1} = \frac{1}{A_2}\frac{\partial \Omega_1}{\partial \alpha_2} - \frac{\Omega_2}{A_1 A_2}\frac{\partial A_2}{\partial \alpha_1}$$

$$= \frac{\Omega_2}{A_1 A_2}\frac{\partial A_2}{\partial \alpha_1} - \frac{\Omega_n}{R_2} - \frac{\Omega_2}{A_1 A_2}\frac{\partial A_2}{\partial \alpha_1} = -\frac{\Omega_n}{R_2},$$

so that we find

$$\tau^* = \tau_1 + \frac{\omega_2}{R_1} = \frac{\Omega_n}{R_1} - \frac{\Omega_n}{R_1} = 0$$

or

$$\tau^* = \tau_2 + \frac{\omega_1}{R_2} = -\frac{\Omega_n}{R_2} + \frac{\Omega_n}{R_2} = 0$$

and finally that

$$\tau = \tau_1 + \tau_2 = \Omega_n\left(\frac{1}{R_1} - \frac{1}{R_2}\right) \neq 0.$$

Thus, to sum up, we have

$$\varepsilon_1^0 = \varepsilon_2^0 = 0, \quad \kappa_1 = \kappa_2 = 0,$$

$$\omega = \omega_1 + \omega_2 = 0, \quad \tau^* = 0,$$

and (j)

$$\tau = \tau_1 + \tau_2 = \Omega_n\left(\frac{1}{R_1} - \frac{1}{R_2}\right).$$

CHAPTER IV

Membrane Shells

IV-1. GENERAL FORMULATION OF MEMBRANE THEORY

We consider now the class of shells known as *membrane shells*, or *membranes*. In these shells, the state of stress is largely a membrane state, because we set

$$M_1 = M_2 = M_{12} = M_{21} = 0. \tag{113}$$

Then equations (82d, e) imply the not unexpected results

$$Q_1 = Q_2 = 0 \tag{114}$$

while the remaining equations of equilibrium take the form

$$\frac{\partial(N_1 A_2)}{\partial\alpha_1} + \frac{\partial(N_{21} A_1)}{\partial\alpha_2} + N_{12}\frac{\partial A_1}{\partial\alpha_2} - N_2\frac{\partial A_2}{\partial\alpha_1} = 0 \tag{115a}$$

$$\frac{\partial(N_{12} A_2)}{\partial\alpha_1} + \frac{\partial(N_2 A_1)}{\partial\alpha_2} + N_{21}\frac{\partial A_2}{\partial\alpha_1} - N_1\frac{\partial A_1}{\partial\alpha_2} = 0 \tag{115b}$$

$$\frac{N_1}{R_1} + \frac{N_2}{R_2} + q_n = 0. \tag{115c}$$

The boundary conditions at an edge α_n = const. are simply

$$N_n = 0 \quad \text{or} \quad \delta u_n = 0$$

$$N_{nt} = 0 \quad \text{or} \quad \delta u_t = 0. \tag{116}$$

Note that for this membrane theory, there are *no* transverse boundary conditions,[6] i.e., no conditions involving $w(\alpha_1, \alpha_2)$ or its derivatives. Also, note that in view of the constitutive law (103), $N_{12} = N_{21}$, and so equations (115) constitute three equations in terms of the three unknowns N_1, N_2 and N_{12}. Thus, according to our theory, membrane shells (stresses) are statically determinate structures!

Once the stresses are calculated, the strains are determined by the inverse of the relations (103),

$$\varepsilon_1^0 = \frac{1}{Eh}(N_1 - \nu N_2), \quad \varepsilon_2^0 = \frac{1}{Eh}(N_2 - \nu N_1)$$

$$\omega = \frac{2(1+\nu)}{Eh} N_{12} = \frac{N_{12}}{Gh} \tag{117}$$

and then the displacements can be determined from equations (66), i.e.,

$$\frac{1}{A_1}\frac{\partial u_1}{\partial \alpha_1} + \frac{u_2}{A_1 A_2}\frac{\partial A_1}{\partial \alpha_2} + \frac{w}{R_1} = \varepsilon_1^0 \tag{118a}$$

$$\frac{1}{A_2}\frac{\partial u_2}{\partial \alpha_2} + \frac{u_1}{A_1 A_2}\frac{\partial A_2}{\partial \alpha_1} + \frac{w}{R_2} = \varepsilon_2^0 \tag{118b}$$

$$\frac{A_2}{A_1}\frac{\partial}{\partial \alpha_1}\left(\frac{u_2}{A_2}\right) + \frac{A_1}{A_2}\frac{\partial}{\partial \alpha_2}\left(\frac{u_1}{A_1}\right) = \omega. \tag{118c}$$

Equations (115)–(118) form a complete set for the solution of the boundary value problems of membrane shells.

For a general (membrane) shell of revolution, recall from equations (38), (39), (42) that

$$\alpha_1 = \phi, \; \alpha_2 = \theta, \; A_1 = R_1, \; A_2 = R_0, \; \frac{dR_0}{d\phi} = R_1 \cos \phi$$

$$(ds)^2 = R_1^2(d\phi)^2 + R_0^2(d\theta)^2. \tag{119}$$

Let $r \equiv R_0$, $r_\phi \equiv R_1$. Then our equilibrium equations become

$$\frac{\partial(rN_\phi)}{\partial \phi} + r_\phi \frac{\partial N_{\phi\theta}}{\partial \theta} - N_\theta r_\phi \cos \phi + q_\phi r r_\phi = 0$$

[6] This is a condition that no longer applies for a *nonlinear* membrane theory.

$$\frac{\partial(rN_{\phi\theta})}{\partial\phi}+r_\phi\frac{\partial N_\theta}{\partial\phi}+N_{\theta\phi}r_\phi\cos\phi+q_\theta rr_\phi=0 \tag{120}$$

$$\frac{N_\phi}{r_\phi}+\frac{N_\theta}{r_\theta}+q_n=0$$

where $r_\theta = r/\sin\phi = R_o/\sin\phi \equiv R_2$. The kinematic conditions (118) become

$$\frac{1}{r_\phi}\left(\frac{\partial u_\phi}{\partial\phi}+w\right)=\varepsilon_\phi^0,\quad \frac{1}{r}\left(\frac{\partial u_\theta}{\partial\phi}+u_\phi\cos\phi+w\sin\phi\right)=\varepsilon_\theta^0$$

$$\frac{1}{r_\phi}\frac{\partial u_\theta}{\partial\phi}-\frac{1}{r}\left(u_\theta\cos\phi-\frac{\partial u_\phi}{\partial\theta}\right)=\omega=\gamma_{\phi\theta}^0. \tag{121}$$

IV-2. SHELLS OF REVOLUTION WITH STRAIGHT GENERATORS

We will specialize our results for shells of revolution with *straight generators*. Thus, we consider cones and cylinders, i.e., the two types of shell where the meridional curve is a straight line. Then, we introduce the variable x such that

$$\lim_{r_\phi\to\infty}(r_\phi d\phi)=dx. \tag{122}$$

Also, it is useful to introduce a coordinate s such that $dy = rd\theta$, so that when we look down at any parallel "circle" plane (Fig. 6) for a cylindrical membrane, we have

$$\frac{\partial N_x}{\partial x}+\frac{\partial N_{xy}}{\partial y}=0,\quad \frac{\partial N_{xy}}{\partial x}+\frac{\partial N_y}{\partial y}=0$$

$$N_y+q_n r=0 \tag{123}$$

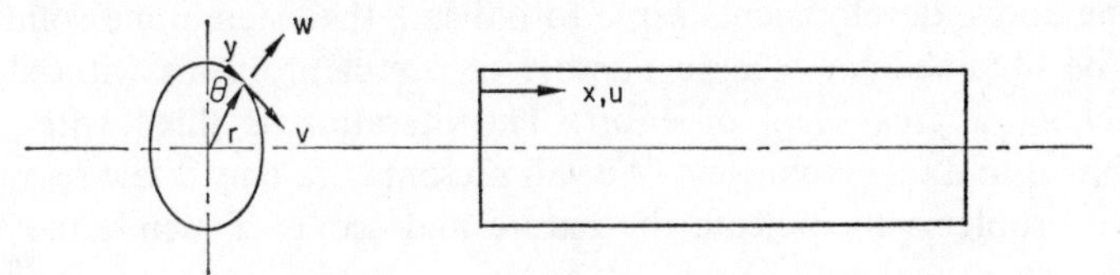

FIG. 6.

and

$$\frac{\partial u}{\partial x} = \varepsilon_x^0, \quad \frac{\partial v}{\partial y} + \frac{w}{r} = \varepsilon_y^0, \quad \gamma_{xy}^0 = \frac{\partial u}{\partial y} + \frac{\partial v}{\partial x}. \tag{124}$$

Consider a modified version of (123) with applied surface loads q_x, q_y:

$$\frac{\partial N_x}{\partial x} + \frac{\partial N_{xy}}{\partial y} + q_x = 0, \quad \frac{\partial N_{xy}}{\partial x} + \frac{\partial N_y}{\partial y} + q_y = 0. \tag{123'}$$

Then it follows that

$$\begin{aligned} N_y &= -q_n r \\ N_{xy} &= -\int \left(q_y + \frac{\partial N_y}{\partial y} \right) dx + f_1(y) \\ N_x &= -\int \left(q_x + \frac{\partial N_{xy}}{\partial y} \right) dx + f_2(y). \end{aligned} \tag{125}$$

Similarly we can get the displacements

$$\begin{aligned} u &= \int \frac{1}{Eh} (N_x - \nu N_y) dx + f_3(y) \\ v &= \int \left(\frac{N_{xy}}{Gh} - \frac{\partial u}{\partial y} \right) dx + f_4(y) \\ w &= r \left[\frac{1}{Eh} (N_y - \nu N_x) - \frac{\partial v}{\partial y} \right]. \end{aligned} \tag{126}$$

It is clear, since f_1, f_2, f_3, f_4 depend only on y, that boundary conditions can be applied at edges $x = \text{const.}$ only.

The above developments serve to indicate that membrane solutions can be obtained for a large number of useful problems without—in principle—a great deal of effort. The literature is filled with such problems and their solutions. We will present here only a few representative problems to indicate the nature and scope of membrane shell usage.

The problems we shall solve are all of axisymmetric loading of shells of revolution. For such axisymmetric conditions, we can stipulate that

$$\frac{\partial}{\partial\theta} = 0, \quad u_\theta = 0, \quad N_{\phi\theta} = 0, \quad q_\theta = 0. \tag{127}$$

Of course this excludes the (axisymmetric) torsion problem. Then the equations of equilibrium (120) reduce to

$$\frac{d}{d\phi}(rN_\phi) - N_\theta r_\phi \cos\phi + q_\phi r r_\phi = 0$$

$$\frac{N_\phi}{r_\phi} + \frac{N_\theta}{r_\theta} + q_n = 0. \tag{128}$$

We can eliminate N_θ between these two equations to find

$$\frac{d}{d\phi}(rN_\phi) + \left(q_n r_\theta + \frac{r_\theta}{r_\phi} N_\phi\right) r_\phi \cos\phi + q_\phi r r_\phi = 0$$

or

$$\frac{d}{d\phi}(rN_\phi) + r_\theta \cos\phi\, N_\phi = -q_\phi r r_\phi - q_n r_\theta r_\phi \cos\phi\,.$$

Now recall the relations $r_\theta = r/\sin\phi$, $(dr/d\phi) = r_\phi \cos\phi$, so that we have

$$\frac{d}{d\phi}(r_\theta \sin\phi N_\phi) + r_\theta \cos\phi\, N_\phi = -[q_\phi r_\phi r_\theta \sin\phi + q_n r_\phi r_\theta \cos\phi]\,.$$

If we multiply by $\sin\phi$ we can rewrite this equation as

$$\frac{d}{d\phi}[r_\theta \sin\phi N_\phi \cdot \sin\phi] = -[q_\phi r_\phi r_\theta \sin^2\phi + q_n r_\phi r_\theta \sin\phi\cos\phi]. \tag{129}$$

Thus we can integrate equation (129) to yield

$$N_\phi = \frac{1}{r_\theta \sin^2\phi}[C_1 - \int (q_\phi r_\phi r_\theta \sin^2\phi + q_n r_\phi r_\theta \sin\phi\cos\phi) d\phi]. \tag{130}$$

In principle we now have determined the stresses.

IV-3. SOME EXAMPLES OF AXISYMMETRIC SHELLS OF REVOLUTION

Consider now a uniform pressure loading on the shell interior, i.e., a *pressurized membrane of revolution* where

$$q_n = -p = \text{const.}, \quad q_\phi = 0. \tag{131}$$

Then in equation (130)

$$\begin{aligned} N_\phi &= \frac{1}{r_\theta \sin^2 \phi} [C_1 + \textstyle\int p r_\phi r_\theta \sin \phi \cos \phi \, d\phi] \\ &= \frac{1}{r_\theta \sin^2 \phi} \left[C_1 + p \int r \frac{dr}{d\phi} \, d\phi \right] \\ &= \frac{1}{r_\theta \sin^2 \phi} \left[C_1 + \frac{pr^2}{2} \right] \end{aligned}$$

$$N_\phi = \frac{C_1}{r_\theta \sin^2 \phi} + \frac{pr_\theta}{2}. \tag{132}$$

If we require now, for a shell complete at the dome, that N_ϕ be finite as $\phi \to 0$, it is clear that we must take $C_1 = 0$. Then it follows from equations (132), (128) that the pressurization introduces a stress state given by

$$N_\phi = \frac{pr_\theta}{2}, \quad N_\theta = pr_\theta \left(1 - \frac{r_\theta}{2r_\phi} \right). \tag{133}$$

We note that in setting $C_1 = 0$ following equation (132), we are assuming that r_θ is finite at $\phi = 0$. We shall consider a case shortly—the pressurized torus—where the finiteness consideration is somewhat more complex. For the record, from equations (133) we can obtain the well-known results for pressurized cylinders and spheres, i.e.,

$$\text{Cylinder:} \quad r_\theta = a, \quad r_\phi \to \infty; \quad N_\phi = \frac{pa}{2}, \quad N_\theta = pa. \tag{134a}$$

$$\text{Sphere:} \quad r_\theta = r_\phi = a; \quad N_\phi = N_\theta = \frac{pa}{2}. \tag{134b}$$

As a further example, we now examine a *spherical* (*membrane*) *cap spinning about the axis of revolution* at constant angular velocity ω (Fig. 7). Now if γ is the specific weight of the shell material, and h is the shell wall thickness, the centripetal force per unit of shell area is given by

$$F = (\omega^2 r)(\text{mass/area}) = (\gamma h/g)(a\omega^2) \sin \phi . \tag{135}$$

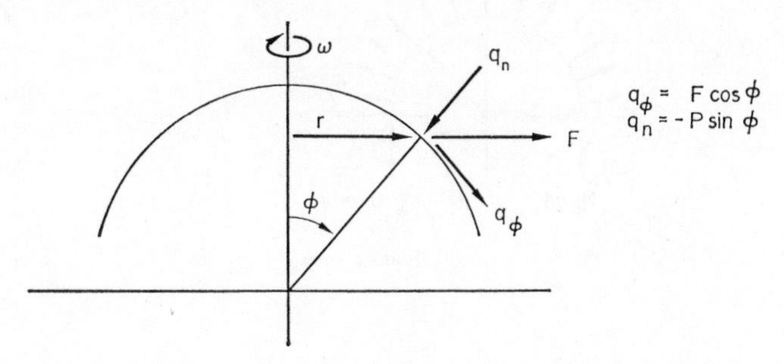

FIG. 7.

Clearly, then,

$$q_n = -(\gamma h/g)(a\omega^2) \sin^2 \phi, \quad q_\phi = (\gamma h/g)(a\omega^2) \sin \phi \cos \phi . \tag{136}$$

Examining the right-hand side of equation (130), together with the results (136), leads to the conclusion that

$$N_\phi = \frac{C_1}{r_\theta \sin^2 \phi} = \frac{C_1}{a \sin^2 \phi} \tag{137}$$

so that for *closed* spherical cap we have, from equations (137), (128), (136),

$$N_\phi = 0, \quad N_\theta = (\gamma h/g)(a^2\omega^2) \sin^2 \phi . \tag{138}$$

At first, this seems a curious result, that is, $N_\phi = 0$. However, reference to Fig. 8 quickly indicates the physical explanation for this result.

We see, in isolating a segment of shell corresponding to a ring of mean radius ($a \sin \phi$), of "thickness" $ad\phi$, that the applied loading always acts normal to the axis of revolution, which is also the axis of spin. Also, this applied loading, which is due to the centripetal acceleration, is uniformly distributed around the circumference, at a rate of ($Fad\phi$) per unit length of circumference. Thus, like a ring under uniform

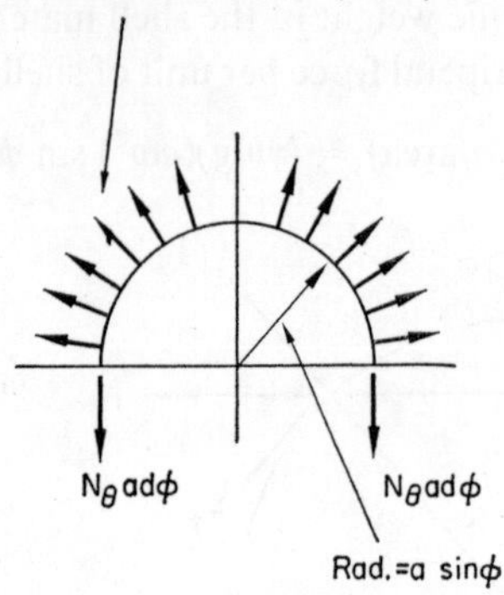

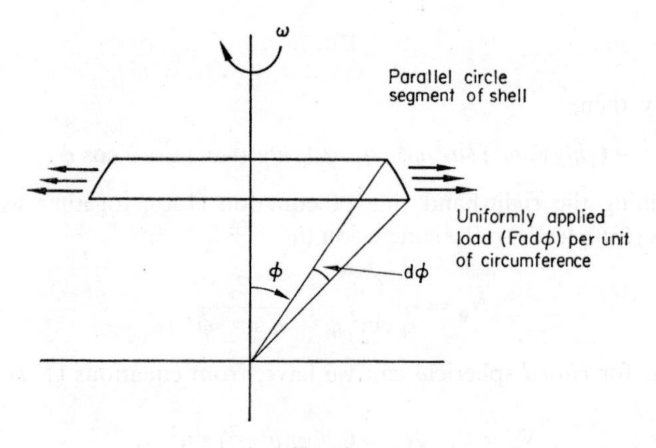

FIG. 8.

pressure, the loading is self-equilibrating. Thus the "ring $ad\phi$" needs no N_ϕ forces to remain in equilibrium. Then by considering a ring of radius $a \sin \phi$, of width/thickness $ad\phi$, and cutting the ring at $\theta = 0$, $\pi/2$ we find

$$2N_\theta(\mathrm{a}d\phi) = (Fad\phi)(2a \sin \phi)$$

which leads to the result (138).

As a final example[7] of finding the stresses in a membrane, we consider a *torus, internally pressurized.* A torus is created by revolving a circle about the axis of revolution (Fig. 9). Here, again, $q_n = -p, q_\phi = q_\theta = 0$. Now for this problem,

$$r_\phi = a, \quad r = d+a \sin \phi, \quad r_\theta = a+d/\sin \phi. \tag{139}$$

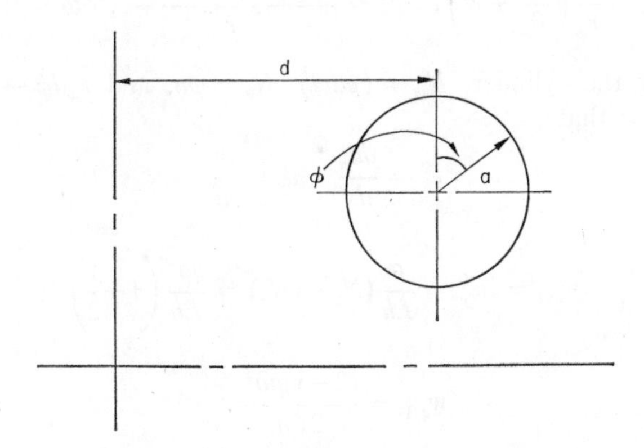

FIG. 9.

Since, for $\phi \to 0$, r_θ is not bounded, we cannot use the previously defined solution (133) for pressurized, axisymmetric membranes. We must go back to equation (132), and substitute from above to get

$$N_\phi = \frac{p(a \sin \phi+d)}{2 \sin \phi}+\frac{C_1}{\sin \phi(a \sin \phi+d)}$$

so that for N_ϕ to be finite as $\phi \to 0, \pi$, we must have $C_1 = -pd^2/2$, and then we get

$$N_\phi = \frac{pa}{2}\left(\frac{a \sin \phi+2d}{a \sin \phi+d}\right), \quad N_\theta = \frac{pa}{2}. \tag{140}$$

Again, for the stress N_θ this time, we have a simple result. Its physical justification is not difficult, and it is left as an exercise.

To close this section on membrane shells, we would like to anticipate some later work by considering the mating of a pressurized cylinder

[7] Two more problems are solved in Appendices IVA and IVB.

with a pressurized hemisphere—to form a head or a cap—of the same radius, subject to the same pressure. We give here the radial displacements for the cylinder and for the cap, so as to compare them. Note first, for axisymmetric deformation, from equations (121):

$$\varepsilon_\phi^0 = \frac{1}{r_\phi}\left(\frac{\partial u_\phi}{\partial_\phi}+w\right), \quad \varepsilon_\theta^0 = \frac{u_\phi \cos\phi + w \sin\phi}{r}, \quad \omega = 0. \tag{141}$$

First, for the cylinder, $N_\phi = (pa/2)$, $N_\theta = pa$, and $r_\phi d\phi \to dx$, and $\phi \to \pi/2$, so that

$$\varepsilon_x^0 = \frac{du_x}{dx}, \quad \varepsilon_y^0 = \frac{w}{a}$$

and thus

$$w_{\text{cyl}} = a\varepsilon_y^0 = \frac{a}{Eh}(N_y - \nu N_x) = \frac{pa^2}{Eh}\left(1-\frac{\nu}{2}\right)$$

or

$$w_{\text{cyl}} = \frac{(2-\nu)pa^2}{2Eh}. \tag{142}$$

In analogous fashion, for the cap—the hemisphere, $r = a \sin\phi$ since $r_\phi = r_\theta = a$, and so

$$\varepsilon_\phi^0 = \frac{1}{a}\left(\frac{du_\phi}{d_\phi}+w\right), \quad \varepsilon_\theta^0 = \frac{u_\phi \cos\phi + w\sin\phi}{a \sin\phi}$$

so that at the cap base $\phi = \pi/2$, and it follows that

$$w_{\text{cap}} = a\varepsilon^0\bigg|_{\phi=\pi/2} = \frac{a}{Eh}(N_\theta - \nu N_\phi)\bigg|_{\phi=\pi/2} = \frac{a}{Eh}(pa)(1-\nu)/2$$

or

$$w_{\text{cap}} = \frac{(1-\nu)pa^2}{2Eh}. \tag{143}$$

Clearly p and a must be the same for both shells. Thus if E, ν, h can be arbitrarily chosen, the radial displacements of cylinder and cap can be matched if

$$\frac{(Eh)_{\text{cyl}}}{(Eh)_{\text{cap}}} = \frac{2-\nu_{\text{cyl}}}{1-\nu_{\text{cap}}}.$$

If the material is the same in both, then

$$\frac{h_{\text{cyl}}}{h_{\text{cap}}} = \frac{2-\nu}{1-\nu}.$$

In general, the mating of two shells causes bending stresses in the neighborhood of the joint; we shall return to this problem in our discussion of shell bending behavior.

APPENDIX IVA

Stresses in a pressurized oval cylinder

WE pose here the problem of determining the stress resultants in a cylindrical membrane shell of oval cross-section, where the radius of curvature is given by

$$\frac{1}{r} = \frac{1}{r_0}\left(1+\xi \cos \frac{4\pi s}{L_0}\right). \tag{a}$$

Here ξ is a non-circularity parameter, L_0 is the oval perimeter, and r_0 is the "equivalent radius" of a circular cylinder of perimeter L_0. The loading considered will be uniform internal pressure. Consider that $u_s = N_x = 0$ at $x = \pm L/2$.

From equations (123), with $y = s$, and with $q_n = -p$, and $q_x = q_y = 0$, we have

$$\frac{\partial N_x}{\partial x}+\frac{\partial N_{xs}}{\partial s} = 0, \quad \frac{\partial N_{xs}}{\partial x}+\frac{\partial N_s}{\partial s} = 0$$

$$N_s = pr = \frac{pr_0}{\left(1+\xi \cos \dfrac{4\pi s}{L_0}\right)}. \tag{b}$$

Since N_s is a function only of s, from equations (125) we deduce that

$$N_{xs} = -x\frac{dN_s}{ds}+f_1(s)$$

$$N_x = \frac{x^2}{2}\frac{d^2N_s}{ds^2}-xf_1'(s)+f_2(s). \tag{c}$$

Hence it follows that

$$N_{xs} = -px\frac{dr}{ds}+f_1(s)$$

$$N_x = \frac{p}{2}(x^2)\frac{d^2r}{ds^2}-xf_1'(s)+f_2(s). \tag{d}$$

From the vanishing of N_x at $x = \pm L/2$ it follows that

$$f_1(s) = c_1 = \text{const}, \quad f_2(s) = -\frac{pL^2}{8}\frac{d^2r}{ds^2}. \tag{e}$$

To complete the problem, we can argue on grounds of symmetry that N_{xs} must be antisymmetric about $x = 0$, and N_{xs} must be zero at $x = 0$. Hence $f_1(s) = c_1 = 0$, and so the stress resultants are

$$N_s = pr = \frac{\text{pr}_0}{\left(1+\xi\cos\frac{4s\pi}{L_0}\right)}$$

$$N_{xs} = -px\frac{dr}{ds} = -\frac{4\pi\xi pr_0}{L_0}\cdot\frac{x\sin(4\pi s/L_0)}{\left(1+\xi\cos\frac{4\pi s}{L_0}\right)^2} \tag{f}$$

$$N_x = \frac{8\pi^2\xi pr_0}{L_0^2}\frac{(x^2-L^{2/4})\left(\xi+\cos\frac{4\pi s}{L_0}+\xi\sin^2\frac{4\pi s}{L_0}\right)}{\left(1+\xi\cos\frac{4\pi s}{L_0}\right)^3}.$$

APPENDIX IVB

Stresses in an ogival dome

WE pose here the problem of determining the stress resultants in an ogival membrane shell of revolution loaded by its own weight. An ogival surface of revolution is formed by rotating a circular arc about an axis which does not correspond to a diameter.

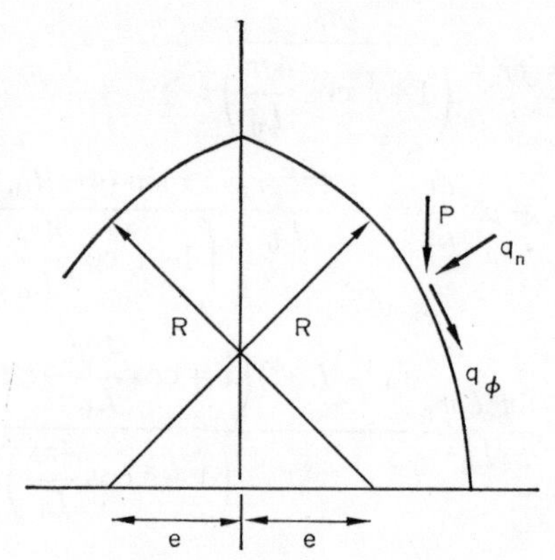

According to the notation defined between equations (119), (120) we have here

$$r = R\sin\phi - e$$

$$r_\phi = R$$

$$r_\theta = r/\sin\phi = R - e/\sin\phi.$$

For the loading we have

$$q_\theta = 0, \quad q_\phi = P\sin\phi, \quad q_n = P\cos\phi.$$

Finally, as the loading is axisymmetric, we may set $N_{\phi\theta} = 0$ and $\partial/\partial\theta = 0$. From equations (128), (130) we then have, using the above values of q_n, q_ϕ, r_ϕ and r_θ:

$$N_\phi = \frac{1}{\sin\phi(R\sin\phi - e)}[c_1 - PR\int[(R\sin\phi - e)\sin^2\phi$$
$$+(R\sin\phi - e)\cos^2\phi]d\phi].$$

We can drop the constant c_1 if we integrate over "definite" limits $\phi_0 \leqq \phi \leqq \phi$ where $\sin\phi_0 = e/R$, or $e = R\sin\phi_0$, so that

$$N_\phi = \frac{PR[R\cos\phi + e\phi]_{\phi_0}^{\phi}}{\sin\phi(R\sin\phi - e)} = \frac{PR^2(\cos\phi - \cos\phi_0) + ePR(\phi - \phi_0)}{\sin\phi(R\sin\phi - e)}$$

or

$$N_\phi = PR\frac{(\cos\phi - \cos\phi_0) + (\phi - \phi_0)\sin\phi_0}{\sin\phi(\sin\phi - \sin\phi_0)}.$$

It quickly follows, then, that

$$N_\theta = -P\cos\phi(R - e/\sin\phi) - N_\phi(1 - e/R\sin\phi)$$
$$= -PR\frac{\cos\phi(\sin\phi - \sin\phi_0)}{\sin\phi} - N_\phi\frac{\sin\phi - \sin\phi_0}{\sin\phi}$$

$$N_\theta = -PR\left[\frac{\cos\phi(\sin\phi - \sin\phi_0)}{\sin\phi} + \frac{(\cos\phi - \cos\phi_0) + (\phi - \phi_0)\sin\phi_0}{\sin^2\phi}\right].$$

CHAPTER V

The Bending of Circular Cylinders

V-1. BASIC RELATIONS AND SIMPLIFICATIONS

To begin our study of *shells with bending*, we will look at the *circular cylinder*. Circular cylindrical shells are of great practicali mportance, so that a great deal of effort has been expended on both theory and practice, giving a variety of materials to draw on.

For a circular cylinder, identifying x as the longitudinal or axial coordinate, $y = R\theta$ as the circumferential, we note that

$$\begin{aligned} &R_1 d\phi \to dx \quad \text{as} \quad R_1 \to \infty; \quad \phi \to \pi/2; \\ &R_0 = R_2 = R; \quad \alpha_1 = x, \alpha_2 = y; \\ &A_1 = 1, A_2 = 1; \quad (ds)^2 = (dx)^2 + (dy)^2. \end{aligned} \tag{144}$$

The *strain–displacement equations* are then

$$\begin{aligned} \varepsilon_x^0 &= \frac{\partial u}{\partial x} & \varepsilon_y^0 &= \frac{\partial v}{\partial y} + \frac{w}{R} \\ \kappa_x &= -\frac{\partial^2 w}{\partial x^2} & \kappa_y &= \frac{1}{R}\frac{\partial v}{\partial y} - \frac{\partial^2 w}{\partial y^2} \\ \gamma_{xy}^0 &= \omega = \frac{\partial u}{\partial y} + \frac{\partial v}{\partial x}, & \tau &= \frac{1}{R}\frac{\partial v}{\partial x} - 2\frac{\partial^2 w}{\partial x \partial y} \end{aligned} \tag{145}$$

The *stress–strain* relations are:

$$\begin{aligned} &N_x = C(\varepsilon_x^0 + \nu\varepsilon_y^0),\ N_y = C(\varepsilon_y^0 + \nu\varepsilon_x^0),\ N_{xy} = Gh\gamma_{xy}^0 \\ &M_x = D(\kappa_x + \nu\kappa_y),\ M_y = D(\kappa_y + \nu\kappa_x),\ M_{xy} = \left(\frac{1-\nu}{2}\right)D\tau. \end{aligned} \tag{146}$$

The *equations of equilibrium* are:

$$\frac{\partial N_x}{\partial x}+\frac{\partial N_{xy}}{\partial y}+q_x = 0 \tag{147a}$$

$$\frac{\partial N_{xy}}{\partial x}+\frac{\partial N_y}{\partial y}+\frac{1}{R}Q_y+q_y = 0 \tag{147b}$$

$$\frac{\partial Q_x}{\partial x}+\frac{\partial Q_y}{\partial y}-\frac{1}{R}N_y-q_n = 0 \tag{147c}$$

$$\frac{\partial M_x}{\partial x}+\frac{\partial M_{xy}}{\partial y}-Q_x = 0 \tag{147d}$$

$$\frac{\partial M_{xy}}{\partial x}+\frac{\partial M_y}{\partial y}-Q_y = 0. \tag{147e}$$

We may eliminate the transverse shear resultants Q_x, Q_y and write equilibrium in the form

$$\begin{aligned}
&\frac{\partial N_x}{\partial x}+\frac{\partial N_{xy}}{\partial y}+q_x = 0\\
&\frac{\partial N_{xy}}{\partial x}+\frac{\partial N_y}{\partial y}+\frac{1}{R}\left(\frac{\partial M_{xy}}{\partial x}+\frac{\partial M_y}{\partial y}\right)+q_y = 0\\
&\frac{\partial^2 M_x}{\partial x^2}+2\frac{\partial^2 M_{xy}}{\partial x\partial y}+\frac{\partial^2 M_y}{\partial y^2}-\frac{1}{R}N_y-q_n = 0.
\end{aligned} \tag{148}$$

Note the resemblance of equations (148) to the equations for the in-plane stretching of plates, and the equation for the bending of rectangular plates. Thus, with $R \to \infty$, the last of equations (148) is the classical plate equation.

By combining equations (145), (146), (148), we can write the equations of equilibrium in terms of the displacements u, v and w. Introduce the dimensionless coordinates $\xi = x/R$, $\eta = y/R$, so that we have

$$\begin{aligned}
L_{11}u+L_{12}v+L_{13}w &= -q_x R^2/C\\
L_{21}u+L_{22}v+L_{23}w &= -q_y R^2/C\\
L_{31}u+L_{32}v+L_{33}w &= -q_n R^2/C
\end{aligned} \tag{149}$$

where the L_{ij} are differential operators defined as

$$
\begin{aligned}
L_{11} &= \frac{\partial^2}{\partial \xi^2} + \frac{1-\nu}{2}\frac{\partial^2}{\partial \eta^2} \\
L_{12} = L_{21} &= \frac{1+\nu}{2}\frac{\partial^2}{\partial \xi \partial \eta} \\
L_{13} = L_{31} &= \nu \frac{\partial}{\partial \xi} \\
L_{22} &= \frac{1-\nu}{2}(1+k^2)\frac{\partial^2}{\partial \xi^2} + (1+k^2)\frac{\partial^2}{\partial \eta^2} \\
L_{23} = L_{31} &= \frac{\partial}{\partial \eta} - \left(\frac{\partial^3}{\partial \xi^2 \partial \eta} + \frac{\partial^3}{\partial \eta^3}\right)k^2 \\
L_{33} = 1 + k^2\nabla^4 &= 1 + k^2\left(\frac{\partial^2}{\partial \xi^2} + \frac{\partial^2}{\partial \eta^2}\right)^2
\end{aligned}
\tag{150}
$$

and where we have introduced a *thickness parameter* k^2 such that

$$
k^2 = \frac{D}{R^2 C} = \frac{1}{12}\left(\frac{h}{R}\right)^2. \tag{151}
$$

It is clear from equation (151) that k^2 gives a measure of the bending effects in the shell, as it represents the ratio of the bending-to-extensional stiffness. Further, for thin shells, $k^2 \ll 1$. Thus we can consider eliminating terms in equations (149), (150) on the basis of the smallness of k^2. In fact simplifications of this type are well known in the literature, and we shall shortly list some of the better-known approximations. It is worth noting that these types of approximations correspond to modifying the strain–displacement relations (145), and, consequently, the equations of equilibrium (147), (148). Thus we are discussing modifications somewhat different than those discussed in Chapter III.

Before proceeding to the simplifications, we also note that the operators L_{ij} are *symmetric*, i.e.,

$$
L_{ij} = L_{ji}. \tag{152}
$$

This is important, for as we shall indicate shortly, symmetry of the differential operators is required for the system of shell equations to satisfy the *Maxwell–Betti Reciprocity Theorem.*

We first note that in the formula for L_{22} we have the term $(1+k^2)$, which is clearly not very different from unity for thin shells where $h/R < 1/10$. Thus we shall simplify these terms. However, to do this consistently we must also inquire as to the origin of the k^2 terms—and, in fact, it is not difficult to verify that they are due to the appearance of Q_y/R in equation (147b), or the second of equations (148). That this is so may be seen from the fact that $k^2 = D/R^2C$, so that in L_{22} the term arises from a bending contribution to the second of equations (148). But these same terms contribute the terms of order k^2 in L_{23}, so that to be consistent we should simplify so that $L_{23} \cong \partial/\partial\xi$.

Now in order to preserve the symmetry properties of the operators L_{ij}, we should also require that $L_{32} \cong \partial/\partial\xi$. This requirement, however, corresponds to deleting the tangential displacement v in the curvature and twist terms, κ_y and τ.

Also, in making the reduction for L_{23}, we note that the simplification is not as easy as setting $1+k^2 \cong 1$ in L_{22}. What we are doing is saying that

$$k^2\left(\frac{\partial^3}{\partial\xi^2\partial\eta}+\frac{\partial^3}{\partial\eta^3}\right) \ll \frac{\partial}{\partial\eta}\,. \tag{153}$$

This is a statement about the nature of function, the derivatives of the function, and their relative rates of changes. Thus a statement about the gradients of the kinematic quantities in the shell is also implied above.

The consequences of dropping the terms on the left-hand side of equation (153) are not obvious, nor are they easy to assess. We will indicate later that the theory thus derived—when compared to more accurate and inclusive theories—is only valid for shorter shells, say $L/R < 4$, or where the deformation pattern involves a large number of circumferential waves, say $n > 4$. We shall return to these points later.

V-2. THE DONNELL, SANDERS AND FLÜGGE EQUATIONS

If we do make the assumptions outlined above, our new simplified shell equations for circular cylinders are termed the *Donnell Equations* (in the United States—in the Soviet Union they are likely to be called the Mushtari–Vlasov equations; internationalists use the name of Mushtari–Vlasov–Donnell equations), and they are:

$$\varepsilon_x^0 = \frac{\partial u}{\partial x}, \quad \varepsilon_y^0 = \frac{\partial v}{\partial y} + \frac{w}{R}, \quad \gamma_{xy}^0 = \frac{\partial u}{\partial y} + \frac{\partial v}{\partial x}$$

$$\kappa_x = -\frac{\partial^2 w}{\partial x^2}, \quad \kappa_y = -\frac{\partial^2 w}{\partial y^2}, \quad \tau = -2\frac{\partial^2 w}{\partial x \partial y} \tag{154}$$

and

$$\frac{\partial N_x}{\partial x} + \frac{\partial N_{xy}}{\partial y} + q_x = 0, \quad \frac{\partial N_{xy}}{\partial x} + \frac{\partial N_y}{\partial y} + q_y = 0$$

$$\frac{\partial^2 M_x}{\partial x^2} + 2\frac{\partial^2 M_{xy}}{\partial x \partial y} + \frac{\partial^2 M_y}{\partial y^2} - \frac{N_y}{R} = q_n \tag{155}$$

and

$$L_{11} = \frac{\partial^2}{\partial \xi^2} + \frac{1-\nu}{2}\frac{\partial^2}{\partial \eta^2}, \quad L_{12} = L_{21} = \frac{1+\nu}{2}\frac{\partial^2}{\partial \xi \partial \eta}$$

$$L_{13} = L_{31} = \nu\frac{\partial}{\partial \xi}, \quad L_{22} = \frac{1-\nu}{2}\frac{\partial^2}{\partial \xi^2} + \frac{\partial^2}{\partial \eta^2} \tag{156}$$

$$L_{23} = L_{32} = \frac{\partial}{\partial \eta}, \quad L_{22} = 1 + k^2\nabla^4.$$

We note that the symmetry of the L_{ij} operators has been preser ed. It will be left as an exercise to show that, using the symmetry of the L_{ij}, the reciprocity theorem for surface loaded shells takes the form[8]

$$\int_0^{L/R}\int_0^{2\pi} (q_x^A u^B + q_y^A v^B - q_n^A w^B)\, d\xi d\eta$$
$$= \int_0^{L/R}\int_0^{2\pi} (q_x^B u^A + q_y^B v^A - q_n^B w^A) d\xi d\eta\,. \tag{157}$$

For an axisymmetric problem, equation (157) is derived in Appendix VC.

As has been often mentioned above, there are many, many shell theories that have been developed. Some of these are based on the

[8] There is a minus sign in front of the radial terms because q_n and w are positive in different directions. Thus in order for all the work terms in equation (157) to be positive, we need that (−) sign.

approximations discussed in Chapter III, some based on types of approximations discussed above, and some a mixture of both.[9]

It is worth noting, for comparison purposes, two sets of equations here—the equations of Sanders and the shell equations derived by Flügge. The equations of Sanders are usually considered the "best" first order shell theory, especially in regard to the internal consistency of assumptions, and satisfaction of reciprocity and the vanishing of all strains for rigid body motions. The equations of Flügge are considered to be the first set of adequate and useful equations for circular cylindrical shells. The strain–displacement relations and the equilibrium equation operators L_{ij} are listed here for both theories.

For *Sanders' equations,*

$$\varepsilon_x^{0s} = \frac{\partial u}{\partial x}, \quad \varepsilon_y^{0s} = \frac{\partial v}{\partial y}+\frac{w}{R}, \quad \gamma_{xy}^{0s} = \frac{\partial v}{\partial x}+\frac{\partial u}{\partial y}$$

$$\kappa_x^S = -\frac{\partial^2 w}{\partial x^2}, \quad \kappa_y^S = \frac{1}{R}\frac{\partial v}{\partial y}-\frac{\partial^2 w}{\partial y^2}, \tag{158}$$

$$\tau^S = \frac{3}{2R}\frac{\partial v}{\partial x}-\frac{1}{2R}\frac{\partial v}{\partial y}-2\frac{\partial^2 w}{\partial x \partial y}$$

and

$$L_{11}^S = \frac{\partial^2}{\partial \xi^2}+\frac{1-\nu}{2}\left(1+\frac{1}{4}k^2\right)\frac{\partial^2}{\partial \eta^2},$$

$$L_{12}^S = L_{21}^S = \left(\frac{1+\nu}{2}-\frac{3(1-\nu)}{8}k^2\right)\frac{\partial^2}{\partial \xi \partial \eta}$$

$$L_{13}^S = L_{31}^S = \nu\frac{\partial}{\partial \xi}+\frac{1-\nu}{2}(k^2)\frac{\partial^3}{\partial \xi \partial \eta^2}, \tag{159}$$

$$L_{22}^S = \frac{1-\nu}{2}\left(1+\frac{9}{4}k^2\right)\frac{\partial^2}{\partial \xi^2}+(1+k^2)\frac{\partial^2}{\partial \eta^2}$$

$$L_{23}^S = L_{32}^S = \frac{\partial}{\partial \eta}-k^2\left(\frac{3-\nu}{2}\frac{\partial^3}{\partial \xi^2 \partial \eta}+\frac{\partial^3}{\partial \eta^3}\right), \quad L_{33}^S = 1+k^2\nabla^4.$$

[9] For a recent discussion of various theories for circular cylinders, see the paper of J. G. Simmonds.

For *Flügge's equations,*

$$\varepsilon_x^F = \frac{\partial u}{\partial x} - z\frac{\partial^2 w}{\partial x^2}, \quad \varepsilon_y^F = \frac{\partial v}{\partial y} + \frac{w}{R+z} - \frac{z}{1+z/R}\frac{\partial^2 w}{\partial y^2}$$

$$\gamma_{xy}^F = \frac{1}{1+z/R}\frac{\partial u}{\partial y} + (1+z/R)\frac{\partial v}{\partial x} - z\left(1+\frac{1}{1+z/R}\right)\frac{\partial^2 w}{\partial x \partial y} \tag{160}$$

and

$$L_{11}^F = \frac{\partial^2}{\partial \xi^2} + \frac{1-\nu}{2}(1+k^2)\frac{\partial^2}{\partial \eta^2}, \quad L_{12}^F = L_{21}^F = \frac{1+\nu}{2}\frac{\partial^2}{\partial \xi \partial \eta}$$

$$L_{13}^F = L_{31}^F = \nu\frac{\partial}{\partial \xi} + k^2\left(\frac{1-\nu}{2}\frac{\partial^3}{\partial \xi \partial \eta^2} - \frac{\partial^3}{\partial \xi^3}\right), \tag{161}$$

$$L_{22}^F = \frac{1-\nu}{2}\left(1+\frac{3}{2}k^2\right)\frac{\partial^2}{\partial \xi^2} + \frac{\partial^2}{\partial \eta^2}, \quad L_{23}^F = L_{32}^F = \frac{\partial}{\partial \eta} - \frac{3-\nu}{2}k^2\frac{\partial^3}{\partial \xi^2 \partial \eta}$$

$$L_{33}^F = (1+k^2) + 2k^2\frac{\partial^2}{\partial \eta^2} + k^2\nabla^4 .$$

We note first of all that in Flügge's strain–displacement relations (160) that we do not have expressions of the form $\varepsilon = \varepsilon^0 + zk$, so that the operators L_{ij}^F reflect in part stress–strain laws where the in-plane stress resultants, for example, are connected to the curvature changes. That is, in contrast to the simple relations (146), in Flügge's equations we will have relations of the form

$$N_x^F = C(\varepsilon_x^0 + \nu\varepsilon_y^0 + Rk^2\kappa_x)$$

$$M_x^F = D\left(\frac{\varepsilon_x^0}{R} + \frac{\nu\varepsilon_y^0}{R} + \kappa_x + \nu\kappa_y\right) \tag{162}$$

and so on. These reflect different degrees of approximation in equations (160) and in the definitions of the stress resultants, of the type discussed at the close of Chapter III.

It is also of interest to note that by appropriately letting $k^2 \to 0$ in the operators L_{ij}^F and L_{ij}^S we can reduce our results to those of Donnell. However, as mentioned earlier, many investigations[10] have demonstrated certain shortcomings of the Donnell equations, even for $k^2 \ll 1$.

[10] See, for example, the papers of J. Kempner, N. J. Hoff, and C. L. Dym, and the text of Kraus.

V-3. THE AXISYMMETRIC, SEMI-INFINITE CYLINDER

At this point we shall present a simple comparison of some of these theories by examining an elementary problem. We will examine the axisymmetric, torsion-free deformation of a semi-infinite cylinder. The cylinder will be subject to a uniform pressure, radially, and a circumferential line load and a moment at the end $x = 0$ (see Fig. 10).

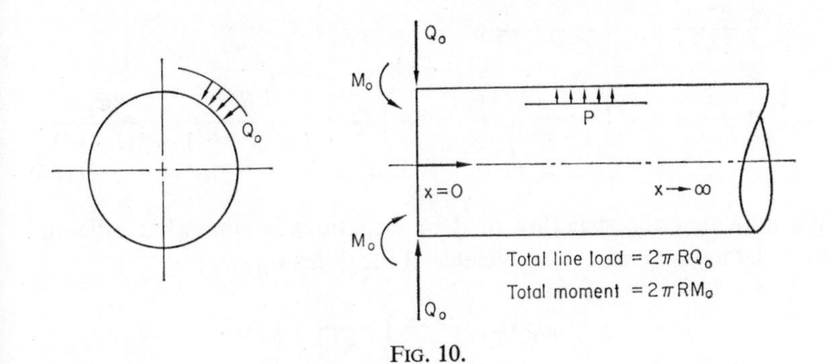

FIG. 10.

To reduce our various sets of equations, we can set $v = 0$, $q_x = q_y = 0$, and all partial derivatives with respect to y are also zero. Then we have, for the various theories:

$$u''_D + \nu w'_D = 0;\; \nu u'_D + w_D + k^2 w_D^{\text{IV}} = \frac{PR^2}{C} \tag{163}$$

$$u''_S + \nu w'_S = 0;\; \nu u'_S + w_S + k^2 w_S^{\text{IV}} = \frac{PR^2}{C} \tag{164}$$

$$u''_F + \nu w'_F - k^2 w'''_F = 0;\; \nu u'_F - k^2 u'''_F + (1+k^2) w_F + k^2 w_F^{\text{IV}} = \frac{PR^2}{C} \tag{165}$$

where we have set $q_n = -P$, the prime denoting ordinary differentiation with respect to ξ, and where the subscripts D, S, F stand for, respectively, Donnell, Sanders, and Flügge. It is clear that, for this case, the Sanders and Donnell equations yield the same simple equation, and so only one of these needs to be dealt with below. We can uncouple the differential equations (163)–(165) to yield equations in the radial displacement only, i.e., from the first equations it is easily seen that

$$u_D'' = -\nu w_D' \quad \text{or} \quad u_D' = -\nu w_D + C_1$$
$$u_F'' = -\nu w_F' + k^2 w_F''' \quad \text{or} \quad u_F' = -\nu w_F + k^2 w_F'' + E_1.$$

Then we can substitute back into equations (163), (165) to find that

$$w_D^{\text{IV}} + \frac{1-\nu^2}{k^2} w_D = \frac{PR^2}{Ck^2} - \frac{\nu}{k^2} C_1$$

$$w_F^{\text{IV}} + \frac{2\nu}{1-k^2} w_F'' + \left(\frac{1}{1-k^2}\right)\left(\frac{1-\nu^2}{k^2}+1\right) w_F = \frac{PR^2}{Ck^2(1-k^2)} - \frac{\nu E_1}{k^2(1-k^2)}. \tag{166}$$

We note that the structure of these equations is somewhat different. First, let us compare the coefficients of the terms w_D, w_F:

$$\frac{1-\nu^2}{k^2} w_D \quad \text{vs.} \quad \frac{1}{1-k^2}\left(\frac{1-\nu^2}{k^2}+1\right) w_F.$$

Recall that for thin shells $h/R < 1/10$, so that $k^2 < 1/1200$, so that obviously the coefficients here are rather the same, for any engineering purpose.

But the Flügge result also has a second order derivative in it, from which we can note that

$$\frac{\text{coefficient of } d^2 w_F/dx^2}{\text{coefficient of } w_F} = \frac{R^2 \text{ coeff. } w_F''(\xi)}{\text{coeff. } w_F(\xi)}$$

$$= \frac{2\nu R^2 k^2}{1+k^2-\nu^2} \cong \frac{\nu h^2}{6(1-\nu^2)} \text{ neglecting } k^2 \ll 1.$$

Thus, except in a region of rapid deformation change, where $w(x)$ and $w'(x)$ are changing a lot over a small distance of the order of the shell thickness, this middle term involving the second derivative will not influence the final result very much.

Let us now write the Donnell equation as

$$\frac{d^4 w_D}{dx^4} + 4\beta^4 w_D = \frac{P}{CR^2k^2}, \quad 4\beta^4 = \frac{1-\nu^2}{R^4k^2} \tag{167}$$

where we are reverting to the dimensional coordinate x, and we are also dropping the constant C_1 which is an extra, useless constant brought in by the elimination process.[11] The solution to equation (167) can be written as, for P = const., remembering the definition of C,

$$w_D = \frac{PR^2}{Eh} + e^{\beta x}(C_2 \cos \beta x + C_3 \sin \beta x)$$
$$+ e^{-\beta x}(C_4 \cos \beta x + C_5 \sin \beta x) . \qquad (168)$$

For a semi-infinite shell, $x \geqq 0$, we must have $C_2 = C_3 = 0$ in order that $w_D(x)$ will be bounded as $x \to \infty$. Thus we take as a solution for a semi-infinite shell,

$$w_D = \frac{PR^2}{Eh} + e^{-\beta x}(C_4 \cos \beta x + C_5 \sin \beta x). \qquad (169)$$

We will determine C_4, C_5 by requiring that at $x = 0$

$$Q_X^D = -\frac{D}{R^3} w_D''' = Q_0, \quad M_X^D = -\frac{D}{R^2} w_D'' = M_0. \qquad (170)$$

Then it follows that, after some algebra,

$$w_D = \frac{PR^2}{Eh} - \frac{1}{2D\beta^3} e^{-\beta x}[Q_0 \cos \beta x + \beta M_0(\cos \beta x - \sin \beta x)] \qquad (171)$$

and the corresponding moment is

$$M_X^D = e^{-\beta x}\left[M_0(\cos \beta x + \sin \beta x) + \frac{Q_0}{\beta} \sin \beta x\right]. \qquad (172)$$

From these results we can now obtain the solution for the deflection of a semi-infinite shell, $0 \leqq x \leqq \infty$, subject only to the internal pressure P, and clamped at the finite edge, i.e., $w(0) = dw(0)/dx = 0$, in addition to $N_x(0) = 0$. This is obtained—as you may readily verify—by forcing Q_0, M_0 to take the values

$$M_0 = -2D\beta^2 \frac{PR^2}{Eh}, \quad Q_0 = 4D\beta^3 \frac{PR^2}{Eh} . \qquad (173)$$

[11] It is not quite useless. By equations (145), (146) taking $C_1 = 0$ implies $u_D' + \nu w_D = RN_x/C = 0$. If the shell end was covered, $C_1 = PR^2/2C$.

These are the support force and moment required to clamp the shell at the origin. Then we obtain for the clamped shell the following radial displacement,[12] moment and shear force

$$w(x) = \frac{PR^2}{Eh}[1-e^{-\beta x}(\sin\beta x+\cos\beta x)] \tag{174a}$$

$$M_x(x) = -D\frac{d^2w}{dx^2} = -2D\beta^2\frac{PR^2}{Eh}e^{-\beta x}[\cos\beta x-\sin\beta x] \tag{174b}$$

$$Q_x = -D\frac{d^3w}{dx^3} = 4D\beta^3\frac{PR^2}{Eh}e^{-\beta x}[\cos\beta x]. \tag{174c}$$

Analogous results for the same problem using Flügge's theory can be found to be

$$w_F(x) = \frac{PR^2}{Eh}\left(\frac{1-\nu^2}{1+k^2-\nu^2}\right)\left[1-e^{-ax}\left(\frac{a}{b}\sin bx+\cos bx\right)\right] \tag{175a}$$

$$\begin{aligned} M_x^F(x) &= \frac{E}{1-\nu^2}\int_{-h/2}^{h/2}(\varepsilon_x^F+\nu\varepsilon_y^F)(1+z/R)z\,dz = -D\left(\frac{d^2w_F}{dx^2}-\frac{1}{R}\frac{du_F}{dx}\right) \\ &= -D\left((1-k^2)\frac{d^2w_F}{dx^2}-\frac{\nu}{R^2}w_F\right) \\ &= \frac{PR^2}{Eh}\left(\frac{1-\nu^2}{1+k^2-\nu^2}\right)[2e^{-ax}(a^2\cos bx-ab\sin bx)-(a^2-b^2)] \end{aligned} \tag{175b}$$

$$Q_x^F(x) = \frac{dM_x^F}{dx} = \frac{PR^2}{Eh}\left(\frac{1-\nu^2}{1+k^2-\nu^2}\right)[(2a^3+2ab^2)(1-k^2)e^{-ax}\cos bx] \tag{175c}$$

where

$$a^2 = \theta^2-\frac{A^2}{2},\quad b^2 = \theta^2+\frac{A^2}{2}$$

$$4\theta^4 = \frac{1+k^2-\nu^2}{R^4k^2(1-k^2)},\quad A^2 = \frac{\nu}{R^2(1-k^2)}. \tag{175d}$$

[12] As we shall be using the Donnell theory almost exclusively beyond this discussion, we drop the subscript D.

Some rather self-explanatory numerical comparisons are displayed below for a shell where $h/R = 1/10$.

x (in.)	βx	R (in.)	$\frac{w_D-1}{w_F-1}$	$\frac{M_D}{M_F}$	$\frac{Q_D}{Q_F}$
0.00	0	10	1.0000000	1.0008736	1.0050077
1.23	0.5	10	1.0002298	0.9897979	1.0043087
2.46	1.0	10	1.0016365	1.0413055	1.0086985
4.92	2.0	10	1.0231466	1.0443516	0.9756399
7.38	3.0	10	0.9889312	1.2033749	0.9906858
0.00	0	50	1.0000000	1.0008736	1.0050077
6.15	0.5	50	1.0002308	0.9897974	1.0043087
12.30	1.0	50	1.0016375	1.0413065	1.0086994
24.60	2.0	50	1.0231543	1.0443563	0.9756352
36.90	3.0	50	0.9889312	1.2034016	0.9906861

V-4. DECAY LENGTHS AND EDGE EFFECTS

It is now useful to use the above Donnell solution to compare some bending results against those already obtained using the membrane theory. We have recorded in equation (142) the radial deflection of a cylindrical membrane, internally pressurized, with capped ends that require $N_x = PR/2$. It is not difficult to deduce that for free ends, where $N_x = 0$, that the (membrane) radial deflection is simply $w_{\text{mem.}} = PR^2/Eh$. Thus equation (175a) can be written in the form

$$\frac{w_{\text{bending}}}{w_{\text{membrane}}} = 1-e^{-\beta x}(\sin \beta x+\cos \beta x)$$

$$= 1-e^{-\beta x}[0(1)]. \tag{176}$$

We now ask the question, how far along the shell must we proceed, in the x direction, for the bending effects introduced at the boundary to be negligible? In other words, how far away from the edge does the state of stress (and deflection) in the shell become substantially that of a membrane? One might also ask the same question about the decay rates of the shear and moments, i.e., examine the relations

$$\frac{M_x(x)}{M_0} = e^{-\beta x}[\cos \beta x - \sin \beta x], \quad \frac{Q_x(x)}{Q_0} = e^{-\beta x} \cos \beta x. \tag{177}$$

From equations (176), (177) we see that the decay of bending effects is exponential. We shall adopt (somewhat arbitrarily) the criterion that $e^{-\pi}=0.043$ is negligible in comparison with unity. Thus we can obtain an estimate for the *critical length*, or *decay length*,[13] x_{cr}, for edge effects as

$$x_{cr} = \frac{\pi}{\beta}\left(\frac{4R^4k^2}{1-\nu^2}\right)^{1/4} = \pi \frac{\sqrt{(Rh)}}{[3(1-\nu^2)]^{1/4}}$$

or

$$x_{cr} = \frac{\pi}{[3(1-\nu^2)]^{1/4}}(Rh)^{1/2}. \tag{178}$$

Thus we see that edge effects—in this problem the influence of Q_0, M_0 (edge loading)—decay over a length proportional to $(Rh)^{1/2}$.

We point out here that the above analysis of characteristic lengths, although conducted for axisymmetric deformation, is in fact correct for asymmetric deformation of circular cylinders, and in general terms for most shells. Thus, for a general shell of revolution, if R_1 or R_2 can be identified as the minimum principal radius, the characteristic length is $\sqrt{(R_1h)}$ or $\sqrt{(R_2h)}$.

As a further exercise, utilizing the simple case of the axisymmetric pressurized shell, we will consider a shell of finite length, capped at the edges by rigid walls. Thus for boundary conditions at $x = \pm L/2$ we will require that

$$N_x = \frac{PR}{2}, \quad w = w' = 0 \quad \text{at} \quad x = \pm L/2. \tag{179}$$

Then, by virtue of previous discussion, the general solution for the Donnell equations is taken as [modifying equation (168)]

$$w(x) = \frac{PR^2}{Eh}\left(1-\frac{\nu}{2}\right) + c_2(\cos \beta x \cosh \beta x + \cos \beta x \sinh \beta x) +$$

[13] x_{cr} is also called a *characteristic length.*

$$+c_4(\cos \beta x \cosh \beta x - \cos \beta x \sinh \beta x)$$
$$+c_3(\sin \beta x \sinh \beta x + \sin \beta x \cosh \beta x) \tag{180}$$
$$+c_5(-\sin \beta x \sinh \beta x + \sin \beta x \cosh \beta x).$$

It is clear that the problem is symmetric about $x = 0$, the center of the shell. Thus we may expect that $w(x)$ is an even—or symmetric—function of x. Thus we may take the solution in the form

$$w(x) = \frac{PR^2}{Eh}\left(1-\frac{\nu}{2}\right)[1+c_2' \cos \beta x \cosh \beta x + c_3' \sin \beta x \sinh \beta x] \tag{181}$$

where c_2', c_3' are the two required constants. After satisfying the boundary conditions (179) we find that

$$w(x) = \frac{PR^2}{Eh}\left(1-\frac{\nu}{2}\right)\left[1-\frac{(\sin \alpha \cosh \alpha + \cos \alpha \sinh \alpha) \cos \beta x \cosh \beta x}{\sin \alpha \cos \alpha + \sinh \alpha \cosh \alpha} + \frac{(\cos \alpha \sinh \alpha - \sin \alpha \cosh \alpha) \sin \beta x \sinh \beta x}{\sin \alpha \cos \alpha + \sinh \alpha \cosh \alpha}\right] \tag{182}$$

where $\alpha = \beta L/2$. Let us now consider the displacement at the center of the shell, i.e., at $x = 0$. Then

$$w(0) = \frac{PR^2}{Eh}\left(1-\frac{\nu}{2}\right)\left[1-\frac{\sin \alpha \cosh \alpha + \cos \alpha \sinh \alpha}{\sin \alpha \cos \alpha + \sinh \alpha \cosh \alpha}\right].$$

If the shell were infinitely long, i.e., with $\alpha \to \infty$, it is not difficult to see that $w(0)$ for $\alpha \to \infty$ approaches the membrane solution (142). We can also note from equation (178) that

$$\alpha = \frac{\beta L}{2} = \frac{\pi}{2}\frac{L}{x_{cr}} \tag{183}$$

so that we can construct the following table for various L/x_{cr}:

L/x_{cr}	α	$\dfrac{w(0)_{\text{bending}}}{w_{\text{membrane}}}$
1	$\pi/2$	0.56
2	π	1.09
3	$3\pi/2$	1.02
4	2π	1.00

We remember that for this problem, with the origin $x = 0$ at the center of the shell, the "disturbances" at the edges occur at a distance $\pm L/2$ from the center. The above table again demonstrates the rapid transition back to the membrane state.

It is also an interesting point to observe that the bending effects occur within a *boundary layer* of "thickness" $\sim \sqrt{(hR)}$ near the disturbance. The mathematical explanation for this stems from the appearance of the small parameter k^2 in front of the highest order derivative in the shell equations. Thus from either of equations (164), (167),

$$k^2 w^{\text{IV}}(\xi) = \frac{1}{12} h^2 R^2 \frac{d^4 w}{dx^4}.$$

In the event that $k^2 \to 0$, then the order of the differential equation for the shell is reduced, thus reducing the number of boundary conditions that $w(x)$ can satisfy. In fact, in the present problem, for the Donnell theory, no boundary conditions on $w(x)$ may be satisfied, as equation (167) immediately dictates $w(x)$ completely, for $k^2 = 0$. For k^2 very small, there will only be a small region around a disturbance—the boundary layer—where $k^2 w^{\text{IV}}(\xi)$ is of the same order of magnitude as w itself in equation (164). The class of problems—particularly the nonlinear problems—that arise when the coefficient of the highest order derivative is quite small, is called the class of *singular perturbation problems*, and it will not be dealt with further here.[14]

We conclude this part of the discussion by pointing out here one of the classical analogies of solid mechanics. The well-known equation that governs the bending of a loaded beam, *on an elastic foundation*, is

$$EI \frac{d^4 w}{dx^4} + kw = q(x)$$

where k is the foundation modulus. Now the differential equation (167) that governs the symmetric deformation of a cylinder can be written as

$$CR^2 k^2 \frac{d^4 w}{dx^4} + C\left(\frac{1-\nu^2}{R^2}\right) w = P$$

[14] See, for example, M. D. Van Dyke, *Singular Perturbation Problems in Fluid Mechanics*, Academic Press, New York, 1965. For applications to shell theory, see the articles by F. John and H. S. Rutten—and the references they cite—in F. Niordson (Ed.), *Theory of Thin Shells*, Springer Verlag, Berlin, 1971.

or

$$D\frac{d^4w}{dx^4}+\frac{Eh}{R^2}w = P.$$

Thus we see that the elastic foundation term in the beam has its analogy in the shell, and it is due to the hoop stresses ($N_y \sim w/R$) induced in the cylindrical shell.

V-5. THE DONNELL EQUATION AND SOME OF ITS SOLUTIONS FOR ASYMMETRIC DEFORMATION

To consider more general problems of deformation of circular cylindrical shells, we return to the equations of equilibrium as expressed in equations (149). Because of the linearity of the operators, L_{ij}, these equations can be uncoupled[15] so as to get a single equation in the radial displacement $w(\xi, \eta)$, and two other equations to be solved for $u(\xi, \eta)$, $v(\xi, \eta)$ after the radial displacement is known. Letting $\Phi_i = -q_iR^2/C$, in symbolic form these equations appear as

$$\begin{aligned}(L_{11}L_{22}L_{33}+2L_{12}L_{13}L_{23}-L_{11}L_{23}^2-L_{22}L_{13}^2-L_{33}L_{12}^2)w \\ = (L_{11}L_{22}-L_{12}^2)\Phi_3+(L_{32}L_{21}-L_{31}L_{22})\Phi_1 \\ +(L_{12}L_{13}-L_{11}L_{23})\Phi_2\end{aligned} \tag{184a}$$

and

$$(L_{11}L_{22}-L_{12}^2)u = (L_{12}L_{23}-L_{22}L_{13})w+L_{22}\Phi_1-L_{12}\Phi_2 \tag{184b}$$

$$(L_{11}L_{22}-L_{12}^2)v = (L_{12}L_{13}-L_{11}L_{23})w+L_{11}\Phi_2-L_{12}\Phi_1. \tag{184c}$$

In the particular case of the Donnell operators, equations (156), we can show that equations (184) become[16]

[15] Another approach to uncoupling the Donnell equations appears in Chapter VI.

[16] In performing these calculations, if we identify three composite operators $L_1 = L_{11}L_{22}-L_{12}^2$, $L_2 = L_{12}L_{23}-L_{22}L_{13}$, $L_3 = L_{12}L_{13}-L_{11}L_{23}$, then the differential equations (184) can be written as

$$(L_{33}L_1+L_{13}L_2+L_{23}L_3)w = L_1\Phi_3+L_2\Phi_1+L_3\Phi_2$$

and

$$L_1u = L_2w+L_{22}\Phi_1-L_{12}\Phi_2$$
$$L_1v = L_3w+L_{11}\Phi_2-L_{12}\Phi_1.$$

$$k^2\nabla^8 w+(1-\nu^2)\frac{\partial^4 w}{\partial\xi^4} = \nabla^4\Phi_3-\left(-\frac{\partial^3\Phi_1}{\partial\xi\partial\eta^2}+\nu\frac{\partial^3\Phi_1}{\partial\xi^3}\right.$$

$$\left.+\frac{\partial^3\Phi_2}{\partial\eta^3}+(2+\nu)\frac{\partial^3\Phi_2}{\partial\xi^2\partial\eta}\right) \quad (185a)$$

$$\nabla^4 u = \frac{\partial^3 w}{\partial\xi\partial\eta^2}-\nu\frac{\partial^3 w}{\partial\xi^3}+\frac{\partial^2\Phi_1}{\partial\xi^2}+\frac{2}{1-\nu}\frac{\partial^2\Phi_1}{\partial\eta^2}-\frac{1+\nu}{1-\nu}\frac{\partial^2\Phi_2}{\partial\xi\partial\eta} \quad (185b)$$

$$\nabla^4 v = -(2+\nu)\frac{\partial^3 w}{\partial\xi^2\partial\eta}-\frac{\partial^3 w}{\partial\eta^3}+\frac{2}{1-\nu}\frac{\partial^2\Phi_2}{\partial\xi^2}+\frac{\partial^2\Phi_2}{\partial\eta^2}-\frac{1+\nu}{1-\nu}\frac{\partial^2\Phi_1}{\partial\xi\partial\eta}. \quad (185c)$$

The set of equations above constitutes an eight-order system, allowing for satisfaction of four boundary conditions at each of the edges ξ = const. and at each of the edges η = const. If we are discussing a shell complete in the circumferential direction, then we require that all quantities be *periodic* in the η coordinate.

The complementary solution can be obtained by considering homogeneous versions of equations (185), in particular the first one,

$$k^2\nabla^8 w^c+(1-\nu^2)\frac{\partial^4 w^c}{\partial\xi^4} = 0. \quad (186)$$

Let us assume that

$$w^c(\xi,\eta) = e^{\lambda\xi}\cos n\eta \quad (187)$$

where $0 \leqq \eta \leqq 2\pi$. Substituting into equation (186) we find that

$$k^2(\lambda^2-n^2)^4+(1-\nu^2)\lambda^4 = 0 \quad (188)$$

or

$$(\lambda^2-n^2)^4+4\beta^4R^4\lambda^4 = 0$$

or

$$(\lambda^2-n^2)^4+4K^4\lambda^4 = 0;\; K^4 = \frac{1-\nu^2}{4k^2}. \quad (189)$$

As this is an eighth-order polynomial, we will obtain a solution with eight unknown constants, corresponding to the eight roots of equation (189). We can, due to the neatness of this polynomial, explicitly solve for the eight roots. This was first done by Hoff, in a slightly different fashion than shown below. Note that equation (189) can be written as:

$$(\lambda^2-n^2)^4 = -4K^4\lambda^4$$

so that it follows that

$$(\lambda^2-n^2)^2 = \pm i2K^2\lambda^2.$$

Since $\pm i = e^{\pm i\pi/2}$

$$\sqrt{(\pm i)} = e^{\pm i\pi/4} = \frac{\sqrt{2}}{2} \pm i\frac{\sqrt{2}}{2}$$

and then

$$\lambda^2-n^2 = \pm\sqrt{(\pm i2K^2\lambda^2)}$$

$$= \pm (K\lambda)\sqrt{2}\left(\frac{\sqrt{2}}{2} \pm i\frac{\sqrt{2}}{2}\right)$$

or

$$\lambda^2-n^2 = \pm K\lambda(1\pm i). \tag{190}$$

Equation (190) represents four quadratic equations, each having two complex roots. We will identify these equations and roots as follows:

$$\lambda_{1,2}\colon\ \lambda^2-n^2 = K\lambda(1+i) \tag{191a}$$

$$\lambda_{3,4}\colon\ \lambda^2-n^2 = -K\lambda(1+i) \tag{191b}$$

$$\lambda_{5,6}\colon\ \lambda^2-n^2 = K\lambda(1-i) \tag{191c}$$

$$\lambda_{7,8}\colon\ \lambda^2-n^2 = -K\lambda(1-i). \tag{191d}$$

Notice that if we apply the conjugation operator to equation (191c) we find that

$$\bar{\lambda}^2-n^2 = K\bar{\lambda}(1+i)$$

so that the conjugates of λ_5, λ_6 satisfy the same equation as λ_1, λ_2. By conjugating equation (191d) we find a similar result for λ_7, λ_8 and λ_3, λ_4, so that we can say

$$\lambda_5 = \bar{\lambda}_1, \quad \lambda_6 = \bar{\lambda}_2, \quad \lambda_7 = \bar{\lambda}_3, \quad \lambda_8 = \bar{\lambda}_4. \tag{192}$$

Further, if we replace λ by $-\lambda$ in equation (191b), it becomes identical to equation (191a). A similar relationship exists between equations (191d) (191c), so we can see that

$$\lambda_3 = -\lambda_1, \quad \lambda_4 = -\lambda_2, \quad \lambda_7 = -\lambda_5, \quad \lambda_8 = -\lambda_6. \tag{193}$$

Combining equations (192), (193) we have

$$\lambda_3 = -\lambda_1, \quad \lambda_4 = -\lambda_2, \quad \lambda_5 = \bar{\lambda}_1, \quad \lambda_6 = \bar{\lambda}_2,$$
$$\lambda_7 = -\bar{\lambda}_1, \quad \lambda_8 = -\bar{\lambda}_2. \tag{194}$$

Thus if we can obtain the two (complex) roots of equation (191a), the remaining six roots can then easily be obtained from equations (194). The two roots in question may be shown to have the form

$$\lambda_1 = \alpha_1 + i\beta_1, \quad \lambda_2 = \alpha_2 + i\beta_2 \tag{195a}$$

where

$$\alpha_1 = \frac{1}{2}\left(K - \frac{n^2}{\Omega_n} - \Omega_n\right) \tag{195b}$$

$$\beta_1 = \frac{1}{2}\left(K - \frac{n^2}{\Omega_n} + \Omega_n\right) \tag{195c}$$

$$\alpha_2 = \frac{1}{2}\left(K + \frac{n^2}{\Omega_n} + \Omega_n\right) \tag{195d}$$

$$\beta_2 = \frac{1}{2}\left(K + \frac{n^2}{\Omega_n} - \Omega_n\right) \tag{195e}$$

$$\Omega_n = \left(-\frac{K^2}{2} + \left[\left(\frac{K^2}{2}\right)^2 + n^4\right]^{1/2}\right)^{1/2}. \tag{195f}$$

Thus, for each value of the *circumferential wave number n*, we will have eight roots $\lambda_{ni} = \lambda_i$ above, so that the complete complementary solution for the radial displacement can be written as

$$w^c(\xi, \eta) = \sum_{n=0}^{\infty} \sum_{j=1}^{8} W_{nj} e^{\lambda_{nj}\xi} \cos n\eta \tag{196}$$

where the W_{nj} are a set of constants to be determined by the satisfaction of boundary conditions.

In order to determine the corresponding axial and circumferential complementary solutions, u^c and v^c, we note first that

$$\nabla^4\left[e^{\lambda\xi}\begin{pmatrix}\sin n\eta\\ \cos n\eta\end{pmatrix}\right] = (\lambda^2-n^2)^2 e^{\lambda\xi}\begin{pmatrix}\sin n\eta\\ \cos n\eta\end{pmatrix}.$$

In view of equations (196), (185b), (185c), and of the above result, we can then easily find that

$$u^c(\xi,\eta) = \sum_{n=0}^{\infty}\sum_{j=1}^{8} U_{nj}e^{\lambda_{nj}\xi}\cos n\eta \tag{197a}$$

$$v^c(\xi,\eta) = \sum_{n=0}^{\infty}\sum_{j=1}^{8} V_{nj}e^{\lambda_{nj}\xi}\sin n\eta \tag{197b}$$

where

$$U_{nj} = -\frac{\lambda_{nj}(\nu\lambda_{nj}^2+n^2)}{(\lambda_{nj}^2-n^2)^2} \tag{198a}$$

$$V_{nj} = \frac{n[(2+\nu)\lambda_{nj}^2-n^2]}{(\lambda_{nj}^2-n^2)^2}. \tag{198b}$$

For the axisymmetric case it is not difficult to show that there are only four non-trivial roots, say $\lambda_{01}, \lambda_{03}, \lambda_{05}, \lambda_{07}$, and they would yield precisely the axisymmetric solution given earlier as equation (168).

It is worthwhile to point out a few features of the above solution. First of all, it is a complementary solution only, and thus does not account for any effects due to surface loading of the shell. If the surface loading vanishes—so that particular solutions to equations (185) are trivial—and if the boundary conditions are homogeneous, the complementary solution must perforce vanish. Similarly, if a particular solution can be found for a given surface loading, and it also satisfies the boundary conditions, then again the complementary solution vanishes.

Further, the solution (196) does represent a radial displacement that is symmetric, around the circumference, with respect to $\eta = 0$. A corresponding—and essentially similar—anti-symmetric solution may be found by substituting $\sin n\eta$ for $\cos n\eta$ in that solution. A solution that is neither symmetric nor anti-symmetric is found by the superposition of these two solutions.

Also, we see that the solution for the in-plane displacements u^c and v^c was made rather easy by the form of the solution (187), and the simplicity of equations (185b, c). This simplicity is a hallmark of the Donnell equations, and accounts in large measure for their frequent use.

A brief discussion of the accuracy of the Donnell equations has already been given, and reference has been made to important early papers of Hoff and Kempner that deal with these questions. A reasonably extensive discussion is also given by Kraus, and we note that comparisons of shell theories still appear frequently in the literature.[17] It is sufficient to note here that the Donnell equations yield adequately accurate results for $n \geqq 4$, and for shells that are not too long, say $L/R \leqq 4$.

As a final indication of the ease of handling and of the wide applicability of the Donnell theory, we consider a shell that is closed in the circumferential direction, and that is simply supported at $x = 0, L$, i.e., at $\xi = 0, L/R$. Thus we require there that

$$N_x = v = w = M_x = 0 \tag{199}$$

or, equivalently,

$$\frac{\partial u}{\partial x} = v = w = \frac{\partial^2 w}{\partial x^2} = 0 \quad \text{at} \quad \xi = 0, L/R. \tag{199a}$$

The loading is only radial, and is assumed to be representable in the double Fourier expansion

$$\Phi_3 = \sum_{m=1}^{\infty} \sum_{n=0}^{\infty} \Phi_{mn} \sin \frac{m\pi x}{L} \cos \frac{ny}{R}$$

or

$$\Phi_3 = \sum_{m=1}^{\infty} \sum_{n=0}^{\infty} \Phi_{mn} \sin \frac{m\pi R\xi}{L} \cos n\eta . \tag{200}$$

This of course implies that the radial pressure distribution $q_n(\xi, \eta)$ is symmetric about $\eta = 0$. We then seek a solution to the

[17] See the papers of J. R. Colbourne, J. H. Williams, and S. H. Iyer and S. H. Simmonds.

differential equation

$$k^2\nabla^8 w+(1-\nu^2)\frac{\partial^4 w}{\partial \xi^4} = \nabla^4\Phi_3$$

$$= \sum_{m=1}^{\infty}\sum_{n=0}^{\infty}\left[\left(\frac{m\pi R}{L}\right)^2+n^2\right]^2 \Phi_{mn}\sin\frac{m\pi R\xi}{L}\cos n\eta .$$

The particular solution is then sought in the form

$$w^p(\xi, \eta) = \sum_{m=1}^{\infty}\sum_{n=0}^{\infty} W_{mn}\sin\frac{m\pi R\xi}{L}\cos n\eta \tag{201}$$

which yields the Fourier coefficients

$$W_{mn} = \frac{\left[\left(\frac{m\pi R}{L}\right)^2+n^2\right]^2\Phi_{mn}}{k^2\left[\left(\frac{m\pi R}{L}\right)^2+n^2\right]^4+(1-\nu^2)\left(\frac{m\pi R}{L}\right)^4}\,. \tag{202}$$

By examination of equations (185b, c) with $\Phi_1 = \Phi_2 = 0$, we see that for the in-plane displacements we have

$$u^p(\xi, \eta) = \sum_{m=1}^{\infty}\sum_{n=0}^{\infty} U_{mn}\cos\frac{m\pi R\xi}{L}\cos n\eta \tag{203a}$$

$$v^p(\xi, \eta) = \sum_{m=1}^{\infty}\sum_{n=0}^{\infty} V_{mn}\sin\frac{m\pi R\xi}{L}\sin n\eta \tag{203b}$$

where

$$U_{mn} = \frac{1}{\Delta_{mn}}\left(\frac{m\pi R}{L}\right)\left[\nu\left(\frac{m\pi R}{L}\right)^2-n^2\right]\Phi_{mn} \tag{203d}$$

$$V_{mn} = \frac{-1}{\Delta_{mn}}(n)\left[(2+\nu)\left(\frac{m\pi R}{L}\right)^2+n^2\right]\Phi_{mn} \tag{203d}$$

and where Δ_{mn} represents the denominator of equation (202).

Ordinarily, having a particular solution in hand, we would now seek a complementary solution, superpose the two, and then proceed to satisfy the boundary conditions (199a). However, it is not difficult to verify for this case that the solution (201), (202), (203) does satisfy the boundary conditions already, so that no complementary solution is required! The reader may recall that this solution is analogous to the classical Navier solution of the theory of plates.

At this point we shall conclude our discussion of the bending of circular cylindrical shells. We have by no means exhausted the topic. Rather we have indicated some different shell theories, some relations between membrane shells and bending effects, and some solution possibilities. Solutions to many shell problems appear in the current technical literature, as well as in the texts and monographs of Flügge, Novozhilov, Timoshenko and Woinowsky-Krieger, Goldenveizer, Vlasov, and Kraus.[18]

[18] See the Bibliography given at the end of the text.

APPENDIX VA

Cylinders with variable wall thickness

WE wish to present here a simple treatment of axisymmetric bending of cylinders of variable wall thickness. For the Donnell theory, the equations of equilibrium would be

$$\frac{dN_x}{d_x}+q_x = 0, \quad \frac{d^2M_x}{dx^2}-\frac{N_y}{R}-q_n = 0 \tag{a}$$

where, with the thickness h being a function of the axial coordinate, $h(x)$,

$$N_x = \frac{Eh}{1-\nu^2}\left(\frac{du}{dx}+\nu\frac{w}{R}\right), \quad N_y = \frac{Eh}{1-\nu^2}\left(\frac{w}{R}+\nu\frac{du}{dx}\right) = \text{const.} \tag{b}$$

$$M_x = -\frac{Eh^3}{12(1-\nu^2)}\frac{d^2w}{dx^2} = -D\frac{d^2w}{dx^2}.$$

If we considered only radial loading, i.e., take $q_x = 0$, then N_x will be a constant. We consider here those problems where $N_x = 0$ for simplicity, so that $du/dx = -\nu w/R$, and so

$$N_y = \frac{Eh}{1-\nu^2}\left(\frac{w}{R}-\nu^2\frac{w}{R}\right) = \frac{Eh}{R}w = \frac{(1-\nu^2)C}{R}w. \tag{c}$$

Combining equations (a), (b), (c) we obtain, for the radial equilibrium condition,

$$\frac{d^2}{dx^2}\left(D\frac{d^2w}{dx^2}\right)+\frac{(1-\nu^2)C}{R^2}w = -q_n \tag{d}$$

or

$$\frac{d^2}{dx^2}\left(h^3 \frac{d^2w}{dx^2}\right)+\frac{12(1-\nu^2)}{R^2}\,hw = -\frac{12(1-\nu^2)q_n}{E}. \tag{e}$$

APPENDIX VB

Influence coefficients for the axisymmetric cylinder

IT is the purpose of this Appendix to introduce the idea of an *influence coefficient*, and its application to shell analysis. From the solution given for an edge loaded shell, also internally pressurized with capped ends, we can obtain the deflection and rotation at the shell edge as [see equation (171)]

$$w(0) = \frac{PR^2}{Eh}\left(1-\frac{\nu}{2}\right)-\left(\frac{1}{2\beta^3 D}\right)Q_0-\left(\frac{1}{2\beta^2 D}\right)M_0$$

$$\beta_x(0) = -\frac{dw(0)}{dx} = -\left(\frac{1}{2\beta^2 D}\right)Q_0-\left(\frac{1}{\beta D}\right)M_0.$$

Notice first of all that

$$\left.\frac{w(0)}{M_0}\right|_{P=Q_0=0} = \left.\frac{\beta_x(0)}{Q_0}\right|_{P=M_0=0}$$

which is a statement of the Maxwell–Betti reciprocal theorem.

The coefficients above, e.g.,

$$\frac{R^2}{Eh}\left(1-\frac{\nu}{2}\right),\ \left(\frac{1}{\beta D}\right),$$

represent the response of the structure to a given type of loading. Noting that β can be written as

$$\beta = -\frac{c}{Rh},\quad c^4 = 3(1-\nu^2)$$

the above equations can be written as

$$w(0) = \frac{(2-\nu)R^2}{2Eh}P - \frac{2c}{E}\left(\frac{R}{h}\right)^{3/2} Q_0 - \frac{2c^2}{E}\left(\frac{R}{h^2}\right)M_0$$

$$\beta_x(0) = -\frac{2c^2}{E}\left(\frac{R}{h^2}\right)Q_0 - \frac{4c^3}{E}\frac{R^{1/2}}{h^{5/2}}M_0.$$

Then the coefficients of P, Q_0, and M_0 are termed *influence coefficients*.

APPENDIX VC

The Maxwell–Betti theorem

IT is the purpose of this Appendix to derive the Maxwell–Betti reciprocal theorem for the axisymmetric Donnell equations. Let there be two sets of displacements and loads, A and B, such that

$$u^A, w^A \quad \text{correspond to} \quad q_x^A, q_n^A$$

$$u^B, w^B \quad \text{correspond to} \quad q_x^B, q_n^B.$$

Now we form the expression [see equations (149)]

$$I = 0 = \int_0^{L/R} \left[\left(L_{11}u^A + L_{13}w^A + \frac{q_x^A R^2}{C} \right) u^B - \left(L_{31}u^A + L_{33}w^A + \frac{q_n^A R^2}{C} \right) w^B \right] d\xi$$

$$= \int_0^{L/R} \left[\left(\frac{d^2u^A}{d\xi^2} + \nu \frac{dw^A}{d\xi} + \frac{q_x^A R^2}{C} \right) u^B - \left(\nu \frac{du^A}{d\xi} + w^A + k^2 \frac{d^4w^A}{d\xi^4} + \frac{q_n^A R^2}{C} \right) w^B \right] d\xi$$

where we have used the axisymmetric reductions of equations (156) to obtain the second formula. We have subtracted the radial equilibrium portion from the axial portion so as to preserve the positive work convention, as was pointed out on p. 82. By some straightforward integrations-by-parts we may then find that

$$I = \left[\left(\frac{du^A}{d\xi} + \nu w^A \right) u^B - k^2 \frac{d^3w^A}{d\xi^3} w^B + k^2 \frac{d^2w^A}{d\xi^2} \frac{dw^B}{d\xi} \right]_0^{L/R} +$$

$$+\int_0^{L/R}\left[\frac{q_x^A R^2}{C}u^B-\frac{q_n^A R^2}{C}w^B-\left(\frac{du^A}{d\xi}+\nu w^A\right)\frac{du^B}{d\xi}\right.$$
$$\left.-\left(\nu\frac{du^A}{d\xi}+w^A\right)w^B-k^2\frac{d^2w^A}{d\xi^2}\frac{d^2w^B}{d\xi^2}\right]d\xi=0.$$

Continuing the integrations-by-parts we have

$$I=\left[\left(\frac{du^A}{d\xi}+\nu w^A\right)u^B-k^2\frac{d^3w^A}{d\xi^3}w^B+k^2\frac{d^2w^A}{d\xi^2}\frac{dw^B}{d\xi}\right.$$
$$\left.-\frac{du^B}{d\xi}u^A-\nu u^A w^B-k^2\frac{dw^A}{d\xi}\frac{d^2w^B}{d\xi^2}+k^2\frac{d^3w^B}{d\xi^3}w^A\right]_0^{L/R}$$
$$+\int_0^{L/R}\left[\frac{q_x^A R^2}{C}u^B-\frac{q_n^A R^2}{C}w^B+\frac{d^2u^B}{d\xi^2}u^A\right.$$
$$\left.-\nu\frac{du^B}{d\xi}w^A+\nu\frac{dw^B}{d\xi}u^A-w^Aw^B-k^2\frac{d^4w^B}{d\xi^4}w^A\right]d\xi=0.$$

If we now invoke the stress–strain relations (146), we see that the boundary terms can be written as

$$I_{\text{boundary}}=\left[\frac{R}{C}(N_x^A u^B-N_x^B u^A)+\frac{R}{C}(Q_x^A w^B-Q_x^B w^A)\right.$$
$$\left.-\frac{1}{C}\left(M_x^A\frac{dw^B}{d\xi}-M_x^B\frac{dw^A}{d\xi}\right)\right]_0^{L/R}.$$

Thus if both sets of functions, A and B, satisfy the same boundary conditions, whether homogeneous or inhomogeneous, then the boundary term vanishes. Then it follows that

$$I=\int_0^{L/R}\left[\frac{q_x^A R^2}{C}u^B-\frac{q_n^A R^2}{C}w^B+\left(\frac{d^2u^B}{d\xi^2}+\nu\frac{dw^B}{d\xi}\right)u^A\right.$$
$$\left.-\left(\nu\frac{du^B}{d\xi}+w^B+k^2\frac{d^4w^B}{d\xi^4}\right)w^A\right]d\xi=0.$$

Noting the equilibrium equations for the set B we have

$$I=\int_0^{L/R}\left[\frac{q_x^A R^2}{C}u^B-\frac{q_n^A R^2}{C}w^B-\frac{q_x^B R^2}{C}u^A+\frac{q_n^B R^2}{C}w^A\right]d\xi=0.$$

Finally,

$$\int_0^{L/R}(q_x^A u^B - q_n^A w^B)d\xi = \int_0^{L/R}(q_x^B u^A - q_n^B w^A)d\xi .$$

The above result is clearly the axisymmetric counterpart of equation (157). It is not difficult to verify that if the L_{ij} are not symmetric, e.g., let $L_{13} = \partial/\partial\xi$ and $L_{31} = (1+k^2)\partial/\partial\xi$, that the above procedure would not work. That is, if the L_{ij} are not symmetric, the reciprocal theorem does not hold.

CHAPTER VI

Shells of Revolution

VI-1. GENERAL FORMULATION AND UNCOUPLING PROCEDURES

In this chapter we shall examine some features of the bending of general shells of revolution. Again we are making no attempt to be exhaustive in our coverage. For a general *shell of revolution*,

$$\begin{aligned}&\alpha_1 = \phi, \quad \alpha_2 = \theta, \quad A_1 = R_1 = r_\phi, \quad \frac{dr}{d\phi} = r_\phi \cos\phi \\ &A_2 = R_0 = r = r_\theta \sin\phi, \quad R_2 = r_\theta.\end{aligned} \tag{204}$$

Note that for a shell of revolution, the geometric quantities above are independent of the circumferential ("polar") angle θ.

The equations of equilibrium are:

$$\begin{aligned}&\frac{\partial}{\partial\phi}(rN_\phi)+r_\phi\frac{\partial N_{\phi\theta}}{\partial\theta}-N_\theta r_\phi\cos\phi+rQ_\phi+rr_\phi q_\phi = 0\\ &\frac{\partial}{\partial\phi}(rN_{\phi\theta})+r_\phi\frac{\partial N_\theta}{\partial\theta}+N_{\phi\theta}r_\phi\cos\phi+r_\phi Q_\theta\sin\phi+rr_\phi q_\theta = 0\\ &\frac{\partial}{\partial\phi}(rQ_\phi)+r_\phi\frac{\partial Q_\theta}{\partial\phi}-\left(\frac{N_\phi}{r_\phi}+\frac{N_\theta}{r_\theta}\right)rr_\phi-rr_\phi q_n = 0\\ &\frac{\partial}{\partial\phi}(rM_\phi)+r_\phi\frac{\partial M_{\phi\theta}}{\partial\phi}-M_\theta r_\phi\cos\phi-rr_\phi Q_\phi = 0\\ &\frac{\partial}{\partial\phi}(rM_{\phi\theta})+r_\phi\frac{\partial M_\theta}{\partial\phi}+M_{\phi\theta}r_\phi\cos\phi-rr_\phi Q_\theta = 0.\end{aligned} \tag{205}$$

The constitutive relations are:

$$N_\phi = C(\varepsilon_\phi^0 + \nu\varepsilon_\theta^0), \quad N_\theta = C(\varepsilon_\theta^0 + \nu\varepsilon_\phi^0), \quad N_{\phi\theta} = \frac{1-\nu}{2}\cdot C\omega$$

$$M_\phi = D(\kappa_\phi + \nu\kappa_\theta), \quad M_\theta = D(\kappa_\theta + \nu\kappa_\phi), \quad M_{\phi\theta} = \frac{1-\nu}{2}\cdot D\tau. \tag{206}$$

And the kinematic relations are

$$\varepsilon_\phi^0 = \frac{1}{r_\phi}\left(\frac{\partial u_\phi}{\partial\phi} + w\right), \quad \varepsilon_\theta^0 = \frac{1}{r}\left(\frac{\partial u_\theta}{\partial\theta} + u_\phi \cos\phi + w \sin\phi\right)$$

$$\omega = \gamma_{\phi\theta}^0 = \frac{r}{r_\phi}\frac{\partial}{\partial\phi}\left(\frac{u_\theta}{r}\right) + \frac{1}{r}\frac{\partial u_\phi}{\partial\theta}$$

$$\kappa_\phi = \frac{1}{r_\phi}\frac{\partial\beta_\phi}{\partial\phi}, \quad \kappa_\theta = \frac{1}{r}\left(\frac{\partial\beta_\theta}{\partial\theta} + \beta_\phi \cos\phi\right) \tag{207}$$

$$\tau = \frac{r}{r_\phi}\frac{\partial}{\partial\phi}\left(\frac{\beta_\theta}{r}\right) + \frac{1}{r}\frac{\partial\beta_\phi}{\partial\theta}$$

$$\beta_\phi = \frac{1}{r_\phi}\left(u_\phi - \frac{\partial w}{\partial\phi}\right), \quad \beta_\theta = \frac{1}{r_\theta}\left(u_\theta - \frac{\partial w}{\partial\theta}\right).$$

Clearly, solving the system of differential equations posed by equations (205), (206), (207) is a non-trivial task. There are two principal approaches. The first is a direct substitution process leading to three partial differential equations in the dependent displacement variables u, v and w. These equations are then reduced to ordinary differential equations by using Fourier series of the form

$$\{u_\phi, w\} = \sum_{m=0}^{\infty} \{u_{\phi m}(\phi), w_m(\phi)\} \cos m\theta$$

$$u_\theta = \sum_{m=1}^{\infty} u_{\theta m}(\phi) \sin m\theta.$$

These expansions, which yield solutions that are periodic of period 2π and thus satisfy continuity requirements, reduce the problem to solving systems of ordinary differential equations.

Another approach is the introduction of a *stress function F*, so that the governing equations are reduced to a pair of differential equations in the dependent variables w and F. Then, again, Fourier series can be used to reduce the partial differential equations to a pair of ordinary differential equations. As the simplest example of that process, let us quickly refer back to Donnell's equations (154), (155) for the circular cylinder—which is a simple case of a shell of revolution. For $q_x = q_y = 0$, let

$$N_x = \frac{\partial^2 F}{\partial y^2}, \quad N_{xy} = -\frac{\partial^2 F}{\partial x \partial y}, \quad N_y = \frac{\partial^2 F}{\partial x^2} \tag{208}$$

so that the first two equations of equilibrium are identically satisfied. The third equation (155) becomes

$$\frac{\partial^2 M_x}{\partial x^2}+2\frac{\partial^2 M_{xy}}{\partial x \partial y}+\frac{\partial^2 M_y}{\partial y^2}-\frac{1}{R}\frac{\partial^2 F}{\partial x^2} = q_n$$

or, after substitution from equations (146), (154),

$$D\nabla^4 w+\frac{1}{R}\frac{\partial^2 F}{\partial x^2} = -q_n. \tag{209}$$

Thus our three equilibrium equations are reduced to one in two dependent variables. The other equation is developed through compatibility, i.e., as we are resolving the in-plane problem through a stress function, we have to insure that a compatible set of displacements can be found. Thus we see from equations (154) that

$$\frac{\partial^2 \varepsilon_x^0}{\partial y^2}+\frac{\partial^2 \varepsilon_y^0}{\partial x^2}-\frac{\partial^2 \gamma_{xy}^0}{\partial x \partial y} = -\frac{1}{R}\frac{d^2 w}{\partial x^2} \tag{210}$$

and further, that

$$\begin{aligned}\frac{\partial^2 \varepsilon_x^0}{\partial y^2}+\frac{\partial^2 \varepsilon_y^0}{\partial x^2}-\frac{\partial^2 \gamma_{xy}^0}{\partial x \partial y} &= \frac{1}{Eh}\left\{\frac{\partial^2 N_x}{\partial y^2}-\nu\frac{\partial^2 N_y}{\partial y^2}\right.\\ &\quad \left.+\frac{\partial^2 N_y}{\partial x^2}-\nu\frac{\partial^2 N_x}{\partial x^2}-2(1+\nu)\frac{\partial^2 N_{xy}}{\partial x \partial y}\right\}\\ &= \frac{1}{Eh}\nabla^4 F. \end{aligned} \tag{211}$$

So that combining equations (210), (211) we have the compatibility condition

$$\nabla^4 F - \frac{Eh}{R}\frac{\partial^2 w}{\partial x^2} = 0. \tag{212}$$

Thus the system of equations (209), (212) form a system of two partial differential equations for the two variables w and F. We note that they can be uncoupled, i.e.,

$$D\nabla^8 F + \frac{Eh}{R^2}\frac{\partial^4 F}{\partial x^4} = -\frac{Eh}{R}\frac{\partial^2 q_n}{\partial x^2}$$

or

$$D\nabla^8 w + \frac{Eh}{R^2}\frac{\partial^4 w}{\partial x^4} = -\nabla^4 q_n. \tag{213}$$

After suitable non-dimensionalization, equation (213) is seen to be identical (for $q_x = q_y = 0$) with equation (185a). Further, we note that if we let $R \to \infty$ in equations (209), (212) we find

$$D\nabla^4 w = -q_n; \quad \nabla^4 F = 0. \tag{214}$$

The first of these is the classical equation of Sophie Germain for the bending of plates; the second is the biharmonic equation found for two-dimensional planar problems in the theory of elasticity. Thus we have here a simple but graphic illustration of the coupling between in-plane and out-of-plane deformation that arises due to curvature. Indeed, one may examine the kinematic conditions (154) of Donnell theory, and it is reasonably clear that the only coupling between the stretching and bending occurs because of the term w/R in ε_y^0. Such uncouplings are properly found in many aspects of shell analysis, although usually not so simply. We shall endeavor to further illustrate such phenomena in the sequel.

VI-2. THE REISSNER–MEISSNER THEORY OF AXISYMMETRIC SHELLS OF REVOLUTION

As a starting point for our discussion we will consider *axisymmetric deformation of shells of revolution*, so that

$$\frac{\partial}{\partial \theta} = 0, \quad u_\theta = N_{\theta\phi} = M_{\theta\phi} = Q_\theta = q_\theta = 0. \tag{215}$$

Then the equations of equilibrium (205) are reduced to three ordinary differential equations, i.e.,

$$\frac{d}{d\phi}(rN_\phi)-N_\theta r_\phi \cos\phi+rQ_\phi+rr_\phi q_\phi = 0$$

$$\frac{d}{d\phi}(rQ_\phi)-\left(\frac{N_\phi}{r_\phi}+\frac{N_\theta}{r_\theta}\right)rr_\phi-rr_\phi q_n = 0 \tag{216}$$

$$\frac{d}{d\phi}(rM_\phi)-M_\theta r_\phi \cos\phi-rr_\phi Q_\phi = 0.$$

It is now useful, following the work of *H. Reissner* and *E. Meissner* to introduce two new variables[19]

$$U = r_\theta Q_\phi, \quad V = \beta_\phi = \frac{1}{r_\phi}\left(u_\phi-\frac{dw}{d\phi}\right). \tag{217}$$

Now if we consider equilibrium of the portion of the shell above a parallel circle at the angle ϕ (Fig. 11) we see that by summing vertical forces—parallel to the axis of revolution—we obtain the equation

$$2\pi r(N_\phi \sin\phi-Q_\phi \cos\phi)+2\pi\int(q_n \cos\phi+q_\phi \sin\phi)rr_\phi\, d\phi = 0. \tag{218}$$

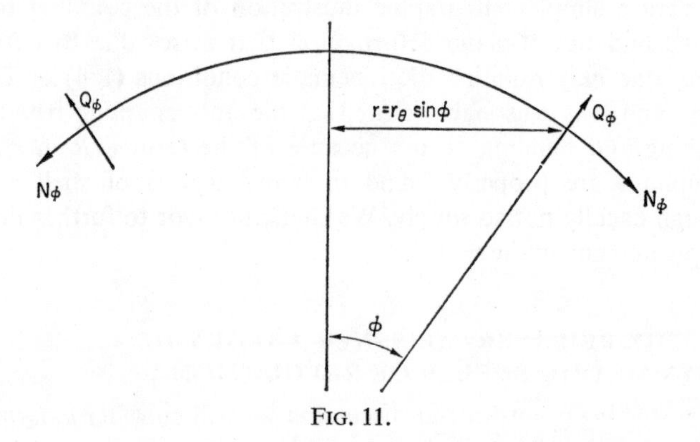

FIG. 11.

[19] For a more recent reference, see Timoshenko and Woinowsky-Krieger. Also, there is available a "Reissner–Meissner" transformation for non-axisymmetric deformation. Details are given in the classic monograph of V. V. Novozhilov.

Note the similarity between equation (218) and its counterpart in membrane theory, equation (130)! For convenience, define

$$Z(\phi) = 2\pi \int (q_n \cos \phi + q_\phi \sin \phi) r r_\phi d\phi \tag{219}$$

so that we have

$$N_\phi = +Q_\phi \cot \phi - \frac{Z}{2\pi r \sin \phi}. \tag{220}$$

Now we combine the first of equations (217) with the above to find

$$N_\phi = +\frac{U \cos \phi}{r} - \frac{Z}{2\pi r \sin \phi}. \tag{221}$$

Then the second of the equations of equilibrium (216) can be solved for the "hoop" resultant N_θ:

$$N_\theta = -\frac{U \cos \phi}{r_\phi \sin \phi} + \frac{1}{r_\phi \sin \phi} \frac{d}{d\phi} (U \sin \phi) + \frac{Z}{2\pi r_\phi \sin^2 \phi} - q_n r_\theta$$

or

$$N_\theta = +\frac{1}{r_\phi} \frac{dU}{d\phi} + \frac{Z}{2\pi r_\phi \sin^2 \phi} - q_n r_\theta. \tag{222}$$

Thus, with equations (221), (222) the membrane stress resultants are now both expressed in terms of the function U (or its equivalent in terms of the shear force, $Q_\phi r_\theta$).

Now we shall manipulate the axisymmetric reductions of the constitutive relations (206) and the kinematic relations (207) to relate the two functions U, V. Since

$$\varepsilon_\phi^0 = \frac{1}{Eh} (N_\phi - \nu N_\theta), \quad \varepsilon_\theta^0 = \frac{1}{Eh} (N_\theta - \nu N_\phi)$$

it follows that

$$\frac{du_\phi}{\partial \phi} + w = \frac{r_\phi}{Eh} (N_\phi - \nu N_\theta)$$

$$u_\phi \cot \phi + w = \frac{r_\theta}{Eh} (N_\theta - \nu N_\phi). \tag{223}$$

Then, eliminating the normal displacement w between equations (223) we find that

$$\frac{du_\phi}{d\phi} - u_\phi \cot\phi = \frac{1}{Eh}\left[N_\phi(r_\phi + \nu r_\theta) - N_\theta(r_\theta + \nu r_\phi)\right]. \tag{224}$$

Now differentiate the second of equations (223)[20]

$$\frac{du_\phi}{d\phi}\cot\phi - \frac{u_\phi}{\sin^2\phi} + \frac{dw}{d\phi} = \frac{1}{Eh}\frac{d}{d\phi}\left[r_\theta(N_\theta - \nu N_\phi)\right]. \tag{225}$$

Now we can eliminate $du_\phi/d\phi$ between equations (224), (225) to obtain

$$\frac{dw}{d\phi} - u_\phi = -r_\phi V = \frac{1}{Eh}\frac{d}{d\phi}\left[r_\theta(N_\theta - \nu N_\phi)\right]$$

$$-\frac{\cot\phi}{Eh}\left[(r_\phi + \nu r_\theta)N_\phi - (r_\theta + \nu r_\phi)N_\theta\right]$$

which in turn yields, after substitution from equations (221), (222):

$$\begin{aligned} EhV = & -\frac{r_\theta}{r_\varphi^2}\frac{d^2U}{d\phi^2} - \frac{1}{r_\phi}\left[\frac{d}{d\phi}\left(\frac{r_\theta}{r_\phi}\right) + \frac{r_\theta}{r_\phi}\cot\phi\right]\frac{dU}{d\phi} \\ & + \frac{1}{r_\phi}\left[\frac{r_\phi}{r_\theta}\cot^2\phi - \nu\right]U \\ & + \frac{1}{r_\phi}\left[\frac{d}{d\phi}(q_n r_\theta^2)\; r_\theta \cot\phi(r_\theta + \nu r_\phi)q_n\right. \\ & \left. - \frac{d}{d\phi}\left(\frac{Z(r_\theta + \nu r_\phi)}{2\pi r_\phi \sin^2\phi}\right) - \frac{Z\cot\phi}{2\pi\sin^2\phi}\left(2\nu + \frac{r_\theta}{r_\phi} + \frac{r_\phi}{r_\theta}\right)\right]. \end{aligned} \tag{226}$$

Thus we have obtained one equation for U, V by manipulating the first two equations of equilibrium. (The second of equations (266) was used explicitly. You may verify that the first was used by multiplying it by $\sin\phi$ and subtracting from it the second, multiplied by $\cos\phi$. This operation yields a differentiated version of equation (218).) A second

[20] At this stage the interested reader can strive for more completeness by allowing $h = h(\phi)$.

relation between U, V is arrived at by noting that the third equation of equilibrium can be written as

$$r_\theta Q_\phi \equiv U = \frac{1}{r_\phi \sin\phi} \frac{d}{d\phi}(rM_\phi) - M_\theta \cot\phi \tag{227}$$

while the stress–displacement conditions for the moments M_ϕ, M_θ and the rotation β_ϕ are

$$M_\phi = D\left(\frac{1}{r_\phi}\frac{d\beta_\phi}{d\phi} + \nu\frac{\beta_\phi \cos\phi}{r}\right) = D\left(\frac{1}{r_\phi}\frac{dV}{d\phi} + \nu\frac{V\cos\phi}{r}\right)$$

$$M_\theta = D\left(\frac{\beta_\phi \cos\phi}{r} + \frac{\nu}{r_\phi}\frac{d\beta_\phi}{d\phi}\right) = D\left(\frac{V\cos\phi}{r} + \frac{\nu}{r_\phi}\frac{dV}{d\phi}\right) \tag{228}$$

so that combining equations (227), (228) yields

$$U = \frac{D}{r_\phi \sin\phi}\frac{d}{d\phi}\left(\frac{r}{r_\phi}\frac{dV}{d\phi} + \nu V\cos\phi\right)$$

$$-D\left(\frac{V\cos^2\phi}{r\sin\phi} + \frac{\nu\cos\phi}{r_\phi \sin\phi}\frac{dV}{d\phi}\right). \tag{229}$$

After some straightforward manipulation, equation (229) can be written in the following compact form:

$$L(V) - \frac{\nu}{r_\phi} V = \frac{U}{D} \tag{230}$$

where L is the differential operator

$$L(---) = \frac{r_\theta}{r_\phi^2}\frac{d^2(---)}{d\phi^2} + \frac{1}{r_\phi}\left[\frac{d}{d\phi}\left(\frac{r_\theta}{r_\phi}\right) + \frac{r_\theta}{r_\phi}\cot\phi\right]\frac{d(---)}{d\phi}$$

$$-\frac{1}{r_\phi}\left[\frac{r_\phi}{r_\theta}\cot^2\phi\right](---). \tag{231}$$

Note that in view of the definition (231), the first relationship (226) between U, V can be written as (here in the homogeneous version)[21]

$$L(U)+\frac{\nu}{r_\phi}U = -EhV. \tag{232}$$

Notice the remarkable symmetry between the differential equations (230), (232). We also see that our forms of these equations allow for direct uncoupling into two individual equations for U, V, i.e.,

$$LL(U)+\nu L\left[\frac{U}{r_\phi}\right]-\frac{\nu}{r_\phi}L(U)-\frac{\nu^2}{r_\phi^2}U = -\frac{Eh}{D}U \tag{233}$$

and

$$LL(V)-\nu L\left(\frac{V}{r_\phi}\right)+\frac{\nu}{r_\phi}L(V)-\frac{\nu^2}{r_\phi^2}V = -\frac{Eh}{D}V. \tag{234}$$

Although equations (230), (232), (233), 234) represent a simplification of the original problem, in view of the nature of the operator L [see equation (231)], the reduced problem is in general not a trivial one. There are a few cases, however, where further immediate simplifications occur.

For a number of shells, e.g., cylinders, cones, spheres, the meridional radius of curvature r_ϕ is a constant. In this instance both equations (233), (234) reduce to the form

$$LL(U)+\mu^4U = 0 \tag{235a}$$

where

$$\mu^4 = \frac{Eh}{D}-\frac{\nu^2}{r_\phi}, \quad r_\phi = \text{const.} \tag{235b}$$

Now equation (235a) can be "factored" into the form

$$[L(U)+i\mu^2U][L(U)-i\mu^2U] = 0\,. \tag{236}$$

[21] Henceforth we shall consider only the homogeneous versions of equations (230), (232). Hence we are able to consider only *edge loading*, except for those instances where the membrane solution is the particular solution, e.g., a pressurized sphere.

It is then an easy matter to show that the four independent solutions of equation (235a) may be obtained from the two pairs of solutions of the second-order differential equations

$$L(U) \pm i\mu^2 U = 0\,. \tag{237}$$

As an illustration, consider once again the *circular cylinder*, for which $\phi = \pi/2$, $r_\theta = R$, $r_\phi d\phi = dx$, while $r_\phi \to \infty$. In this case, then, equations (237) become—after simplifying the operator (231)—

$$R\frac{d^2U}{dx^2} \pm i\mu_c^2 U = 0 \tag{238a}$$

where

$$\mu_c^4 = \frac{Eh}{D} = \frac{12(1-\nu^2)}{h^2} \tag{238b}$$

so that the solutions of equation (238a) may be written as

$$U(x) = \sum_{j-1}^{4} A_j e^{\lambda_j x} \tag{239a}$$

where the λ_j are given by

$$\lambda_j = \pm\left[\frac{\sqrt{2}}{2} \pm i\frac{\sqrt{2}}{2}\right]\frac{[12(1-\nu^2)]^{1/4}}{hR}\,. \tag{239b}$$

In terms of the parameter β defined in equation (167) these roots are simply written as

$$\lambda_j = \pm(1 \pm i)\beta\,. \tag{240}$$

Thus the solution consisting of equations (239a), (240) is seen to be equivalent to that given as equation (168).

For a *sphere* of radius R the equations (237) become

$$\frac{1}{R}\frac{d^2U}{d\phi^2} + \frac{\cot\phi}{R}\frac{dU}{d\phi} - \left(\frac{\cot^2\phi}{R} \mp i\mu^2\right)U = 0\,. \tag{241}$$

Actually, it will be useful for us to give here not just the homogeneous versions, but those that include a uniform (radial) normal pressure,

$q_n = P$. In this instance it is easy to verify that the governing equations are in fact homogeneous, however, with

$$RLQ_\phi + \nu Q_\phi = -Eh\beta_\phi \tag{242a}$$

$$RL\beta_\phi - \nu\beta_\phi = \frac{R^2}{D} Q_\phi \tag{242b}$$

$$RL(\ldots) = \frac{d^2(---)}{d\phi^2} + \cot\phi \frac{d(---)}{d\phi} - \cot^2\phi(---). \tag{242c}$$

What this says is, in fact, that the particular solutions—due to the applied surface loading (here a constant radial pressure)—are such that

$$\beta_\phi^P = Q_\phi^P = 0 \tag{243}$$

which implies that

$$N_\phi^P = -\frac{Z}{2\pi r_\theta \sin^2\phi} = -\frac{PR}{2} \tag{244a}$$

$$N_\theta^P = \frac{Z}{2\pi r_\phi \sin^2\phi} - q_n r_\theta = -\frac{PR}{2} \tag{244b}$$

$$M_\phi^P = M_\theta^P = 0. \tag{244c}$$

Thus the particular solution for uniform radial pressure is just the membrane solution!

The complementary solution for this problem can be obtained from equation (241), repeated here as

$$\frac{d^2Q_\phi}{d\phi^2} + \cot\phi \frac{dQ_\phi}{d\phi} - (\cot^2\phi \mp i\mu_s^2)Q_\phi = 0, \tag{241a}$$

with

$$\mu_s^2 = \left[\frac{Eh}{D} - \frac{\nu^2}{R^2}\right]^{1/2} R = \left[\frac{12(1-\nu^2)R^2}{h^2} - \nu^2\right]^{1/2}.$$

If we introduce a change of variables into equation (241a), namely

$$x = \sin^2\phi, \quad Q_\phi = z\sin\phi \tag{245}$$

then equation (241a) is transformed into a *hypergeometric equation* of the form

$$x^2(x-1)\frac{d^2z}{dx^2}+\left(\frac{5}{2}x-2\right)x\frac{dz}{dx}+(1\mp i\mu_s^2)x\frac{z}{4} = 0. \qquad (246)$$

Solutions to such equations are well known, and are given in terms of infinite series. In the present instance these series converge absolutely for $0 \leqq \phi \leqq \pi/2$, although the rate of convergence is dependent on the ratio R/h, through the parameter μ_s. Some explicit solutions are presented in the texts of Timoshenko and Woinowsky-Krieger and of Kraus. We shall not get into those details here. It is sufficient for our purposes to point out that even for "simple" problems such as the axisymmetric deformation of a sphere, the solutions are found only by effort that is far from trivial.

VI-3. THE GECKELER APPROXIMATION FOR STEEP SHELLS

In view of these remarks we shall present some approximations developed for various classes of shells of revolution, particularly spheres. The classes are two, when the shells are considered either *shallow* or *steep*. In the former case the meridional angle ϕ is rather small, so that the *altitude* of the "cap"—measured along the axis of revolution from the apex to the parallel circle defining the shell base or edge—is small compared to the base diameter. When we speak of steeper shells we will be referring to those where the meridional angle at the edge is near $\phi = \pi/2$. Thus we consider here, among others, hemispheres.

We will consider first the *steeper* shells, and we will develop the *Geckeler approximation*, in particular for spherical shells. We note, though, that it can be developed for more general shells of revolution. Also, we should note that we are particularly interested in edge loading of these steeper shells.

In our discussion of the bending of circular cylinders in Chapter V, we pointed out that bending due to edge loading or edge restrictions occurs only in a small region near the shell edge. Away from this boundary layer, the shell generally behaves as a membrane. In view of this we would argue that only the highest order derivative should be retained in the equations of equilibrium, because in a region of rapid change it will be the most significant term. Also, we note that near $\phi = \pi/2$, $\cot \phi$ is very small. Thus in equation (242c) for the operator L for spherical shells, the argument for retaining only the highest order derivative on the left-hand sides of equations (242a, b) is further buttressed. Thus we shall approximate these equations by

$$\frac{d^2 Q_\phi}{d\phi^2} = -Eh\beta_\phi \tag{247a}$$

$$\frac{d^2 \beta_\phi}{d\phi^2} = \frac{R^2}{D} Q_\phi. \tag{247b}$$

In the uncoupled form we have

$$\frac{d^4 Q_\phi}{d\phi^4} + 4\lambda^4 Q_\phi = 0 \tag{248}$$

where

$$4\lambda^4 = \frac{EhR^2}{D} = \frac{12(1-\nu^2)R^2}{h^2}. \tag{249}$$

We are already familiar with the solution to equation (248) from our work on the axisymmetric bending of circular cylinders. From that work we select as a solution for our present problem

$$Q_\phi(\phi) = (A_1 \cos \lambda\phi + A_2 \sin \lambda\phi)e^{\lambda\phi}. \tag{250}$$

Thus we retain here the solution that increases over the interval $\phi = 0$ to $\phi = \pi/2$, say. In other words, we keep that part of the solution that decays as we move from the shell edge towards the shell apex at $\phi = 0$.

As an example, we consider briefly the deformation of a pressurized hemisphere that is clamped at the parallel circle $\phi = \pi/2$. The appropriate boundary conditions for clamping are

$$w = \beta_\phi = 0 \quad \text{at} \quad \phi = \pi/2. \tag{251}$$

We can write these in another form by noting that (for ϕ close to $\pi/2$, so $w \cong \Delta r_\theta = \varepsilon_\theta^0 r_\theta$)

$$w \cong \varepsilon_\theta^0 R \sin \phi = \frac{R}{Eh}(N_\theta - \nu N_\phi)\sin \phi$$

$$= \left\{\frac{PR^2(1-\nu)}{2Eh} + \frac{R}{Eh}\left[\frac{dQ_\phi}{d\phi} - \nu Q_\phi \cot \phi\right]\right\} \sin \phi \qquad (252)$$

so that the boundary conditions can be written as

$$\beta_\phi = 0, \quad \frac{dQ_\phi}{d\phi} = -\frac{PR}{2}(1-\nu) \quad \text{at} \quad \phi = \pi/2. \qquad (253)$$

Then, after satisfying these boundary conditions, we obtain, for example,

$$N_\phi = \frac{PR}{2}[1 + e^{\lambda\phi} \cot \phi\, (B_1 \cos \lambda\phi + B_2 \sin \lambda\phi)] \qquad (254a)$$

$$M_\phi = -\frac{PD\lambda^2}{Eh} e^{\lambda\phi}[\lambda B_2(\cos \lambda\phi - \sin \lambda\phi) - \lambda B_1(\sin \lambda\phi + \cos \lambda\phi)$$

$$+ \nu \cot \phi(B_2 \cos \lambda\phi - B_1 \sin \lambda\phi)] \qquad (254b)$$

where

$$B_1 = -\frac{1-\nu}{\lambda} e^{-\lambda\pi/2} \cos \lambda\pi/2 \qquad (254c)$$

$$B_2 = -\frac{1-\nu}{\lambda} e^{-\lambda\pi/2} \sin \lambda\pi/2. \qquad (254d)$$

Note that in computing M_ϕ in the above, we have used the complete definition (228), and have not argued that $d\beta_\phi/d\phi \gg \nu\beta_\phi \cot \phi$. This is a practice that is not uniformly ascribed to. For example, in obtaining the second boundary condition (253) from equation (252), there is no problem for $\phi = \pi/2$. What do we do, however, for $\phi = \alpha \cong \pi/2$? In the differential equations we kept only the highest derivatives, and thus it would seem logical to use also the approximations (given here for the sphere)

$$M_\phi \cong \frac{D}{R}\frac{d\beta_\phi}{d\phi}, \quad M_\theta \cong \frac{\nu D}{R}\frac{d\beta_\phi}{d\phi} = \nu M_\phi$$

$$\varepsilon_\theta^0 = \frac{R}{EH}\frac{dQ_\phi}{d\phi} \quad \text{(ignoring any pressure).} \tag{255}$$

We shall use these approximations in our next problems.

For this problem, not so incidentally, both the texts of Kraus and of Timoshenko and Woinowsky-Krieger present numerical results that indicate how excellent an approximation the Geckler solution is to the exact solution of equations (242). The following small table is taken from Kraus's Table 7.1.

$\phi°$	$2N_\theta/R$	$(R/h = 10)$	M_θ/Ph^2	$(R/h = 10)$
	Exact	Geckeler	Exact	Geckeler
90	0.300	0.300	−1.066	−1.059
80	0.514	0.514	−0.061	−0.052
70	0.802	0.807	+0.219	+0.221
60	0.970	0.973	+0.186	+0.178
		$(R/h = 100)$		$(R/h = 100)$
	Exact	Geckeler	Exact	Geckeler
90	0.300	0.300	−10.598	−10.591
80	0.988	0.988	+1.591	+1.583
70	1.010	1.009	−0.092	−0.091
60	0.999	0.999	−0.006	−0.006

Let us consider (Fig. 12) a spherical dome, internally pressurized, and loaded with edge forces and moments as shown. For this problem, the solution (250) is subject to the boundary conditions

$$M_\phi(\alpha) = M_\alpha, \quad N_\phi(\alpha) = \frac{PR}{2} - H_\alpha \cos\alpha. \tag{256}$$

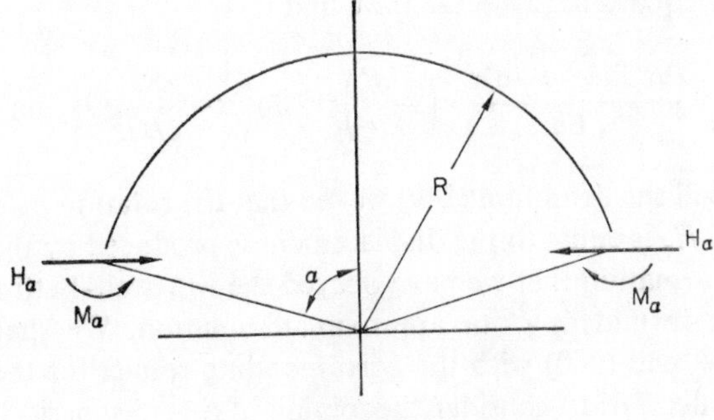

FIG. 12.

In view of equations (220), (255), (247a) these boundary conditions can be written as

$$Q_\phi(\alpha) = -H_\alpha \sin\alpha$$

$$\frac{d^3 Q_\phi(\alpha)}{d\phi^3} = -\frac{EhR}{D} M_\alpha. \tag{257}$$

After satisfying these boundary conditions, the solution can be written as

$$Q_\phi(\phi) = e^{\lambda(\phi-\alpha)}[A_1 \cos\lambda(\phi-\alpha)+(A_1-A_2)\sin\lambda(\phi-\alpha)] \tag{258a}$$

$$\beta_\phi(\phi) = \frac{2\lambda^2}{Eh} e^{\lambda(\phi-\alpha)}[A_1 \sin\lambda(\phi-\alpha)-(A_1-A_2)\cos\lambda(\phi-\alpha)] \tag{258b}$$

where

$$A_1 = -H_\alpha \sin\alpha, \quad A_2 = \frac{EhR}{2D\lambda^3} M_\alpha. \tag{258c}$$

We shall make use of the above results to obtain the *influence coefficients* for a hemispherical head, for the deflection and slope at $\phi = \alpha$. From equation (258b) we easily find that

$$\beta_\phi(\alpha) = \left(\frac{2\lambda^2}{Eh}\right) H_\alpha \sin\alpha + \frac{R}{\lambda D} M_\alpha \tag{259b}$$

while using equations (258), (252) we find

$$w(\alpha) = \frac{PR^2(1-\nu)\sin\alpha}{2Eh} - \left(\frac{2\lambda R}{Eh}\right) H_\alpha \sin^2\alpha - \frac{R^2}{2D\lambda^2} M_\alpha \sin\alpha \qquad (260)$$

If we recall the definition (249) we see that the rotation β_ϕ due to the shear force H_α is equal to the displacement w produced by the moment M_α—a confirmation that we have obeyed the Maxwell–Betti reciprocal theorem in formulating our approximate solution. We shall use the results (259) and (260) with the corresponding results for the cylinder (see Appendix VB) to consider the joining of a hemispherical cap with a circular cylindrical shell. The different radial expansions of these two shells—which we discussed in Chapter IV—will produce *discontinuity stresses* if they are forced to stay together. The details are presented in Appendix VIA.

VI-4. THE REISSNER THEORY FOR SHALLOW SHELLS

As a final topic for this chapter we shall examine the theory of *shallow spherical caps*, which is due to Eric Reissner. Consider the cap drawn in Fig. 13. We postulate the word shallow to mean that the shell height H is less than one-eighth of the base diameter, $2a$. If we introduce a new altitude coordinate z and a new meridional coordinate r, it is clear from Fig. 13 that

$$z = \sqrt{(R^2-r^2)} - (R-H). \qquad (261)$$

At $z = 0$, $r = a$, so that for shallow shells

$$a^2 = 2HR\left(1-\frac{H}{2R}\right) \cong 2HR. \qquad (261a)$$

This follows because $R^2 = (R-H)^2+a^2$, and in this expression if we let $a = \varepsilon H$, with $\varepsilon > 4$ according to Reissner's postulate, then $H/R = (1+\varepsilon^2)^{-1} \ll 1$. Then it also follows that

$$\sin\phi_0 = \frac{a}{R} = \frac{a^2}{aR} = \frac{2H}{a} < \frac{1}{2}, \quad \phi_0 < 30°. \qquad (261b)$$

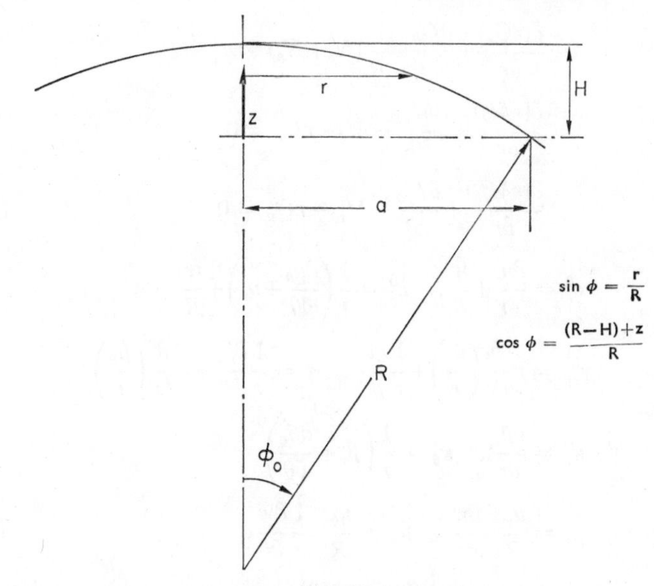

FIG. 13.

Thus we have a number of equivalent conditions, according to which a shell may be taken as shallow, i.e.,

$$\frac{H}{a} < \frac{1}{4};\ \frac{H}{R} < \frac{1}{17};\ \frac{a}{R} < \frac{1}{2};\ \phi_0 < 30^0. \tag{261c}$$

For a shallow spherical cap, we take

$$R_1 = R_2 = r_\phi = r_\theta = R,\ A_1 = R,\ A_2 = r,\ dr \cong Rd\phi. \tag{262}$$

Then the fundamental equations can be listed as (where the subscript ϕ has been changed to r, purely as a matter of notation, without changing the physical meaning)

$$\frac{\partial(rN_r)}{\partial r} + \frac{\partial N_r}{\partial\theta} - N_\theta + \frac{r}{R}Q_r + rq_r = 0$$

$$\frac{\partial(rN_{r\theta})}{\partial r} + \frac{\partial N_\theta}{\partial\theta} + N_{r\theta} + \frac{r}{R}Q_\theta + rq_\theta = 0$$

$$\frac{\partial(rQ_r)}{\partial r}+\frac{\partial Q_\theta}{\partial \theta}-\frac{r}{R}(N_r+N_\theta)-rq_n = 0$$

$$\frac{\partial(rM_r)}{\partial r}+\frac{\partial M_{r\theta}}{\partial \theta}-M_\theta-rQ_r = 0$$

$$\frac{\partial(rM_{r\theta})}{\partial r}+\frac{\partial M_\theta}{\partial \theta}+M_{r\theta}-rQ_\theta = 0 \tag{263}$$

$$\varepsilon_r^0 = \frac{\partial u_r}{\partial r}+\frac{w}{R}, \quad \varepsilon_\theta^0 = \frac{1}{r}\left(\frac{\partial u_\theta}{\partial \theta}+u_r\right)+\frac{w}{R}$$

$$\gamma_{r\theta}^0 = r\frac{\partial}{\partial r}\left(\frac{u_\theta}{r}\right)+\frac{1}{r}\frac{\partial u_r}{\partial \theta}, \quad \tau = \frac{1}{r}\frac{\partial \beta_r}{\partial \theta}+r\frac{\partial}{\partial r}\left(\frac{\beta_\theta}{r}\right)$$

$$\kappa_r = \frac{\partial \beta_r}{\partial r}, \quad \kappa_\theta = \frac{1}{r}\left(\beta_r+\frac{\partial \beta_\theta}{\partial \theta}\right)$$

$$\beta_r = \frac{u_r}{R}-\frac{\partial w}{\partial r}, \quad \beta_\theta = \frac{u_\theta}{R}-\frac{1}{r}\frac{\partial w}{\partial \theta}. \tag{264}$$

By comparing equations (205), (263) we see that we have maintained the relation $R \sin \phi \cong r_\theta \sin \phi \cong r$, and have also put $\cos \phi \cong 1$ in appropriate instances.

In addition to these simplifications, Reissner has argued that for a shallow shell, the influence of the transverse shear forces on the "in-plane" equilibrium equations is small, as is the effect of the in-plane displacements u_r, u_θ on the curvature and twist terms. Therefore the first two equations of equilibrium reduce to

$$\frac{\partial(rN_r)}{\partial r}+\frac{\partial N_{r\theta}}{\partial \theta}-N_\theta+rq_r = 0$$

$$\frac{\partial(rN_{r\theta})}{\partial r}+\frac{\partial N_\theta}{\partial \theta}+N_{r\theta}+rq_\theta = 0 \tag{265}$$

while the new kinematic expressions are

$$\kappa_r = -\frac{\partial^2 w}{\partial r^2}, \quad \kappa_\theta = -\frac{1}{r}\left(\frac{\partial w}{\partial r}+\frac{1}{r}\frac{\partial^2 w}{\partial \theta^2}\right)$$

$$\tau = -2\frac{\partial}{\partial r}\left(\frac{1}{r}\frac{\partial w}{\partial \theta}\right). \tag{266}$$

As we have seen before, equations (265) are counterparts of (in polar coordinates) the plane stress equations of elasticity. Thus, for our purposes, we note that if the surface loads are derivable from a potential Φ,

$$q_r = -\frac{\partial \Phi}{\partial r}, \quad q_\theta = -\frac{1}{r}\frac{\partial \Phi}{\partial \theta} \tag{267}$$

then equations (265) may be identically satisfied if the membrane stress resultants are derivable from the stress function F by

$$N_\theta = \frac{\partial^2 F}{\partial r^2}+\Phi, \quad N_r = \frac{1}{r}\frac{\partial F}{\partial r}+\frac{1}{r^2}\frac{\partial^2 F}{\partial \theta^2}+\Phi$$

$$N_{r\theta} = -\frac{\partial}{\partial r}\left(\frac{1}{r}\frac{\partial F}{\partial \theta}\right). \tag{268}$$

If we are going to use a stress function to satisfy equilibrium in the plane of the shell, then we must insure that u_r, u_θ are single-valued and continuous by deriving and using a compatibility condition. We can eliminate u_r, u_θ from a combination of e_θ^0, ε_r^0, $\gamma_{r\theta}^0$, yielding the result

$$\frac{1}{r^2}\frac{\partial^2 \varepsilon_r^0}{\partial \theta^2}-\frac{1}{r}\frac{\partial \varepsilon_r^0}{\partial r}+\frac{1}{r^2}\frac{\partial}{\partial r}\left(r^2\frac{\partial \varepsilon_\theta^0}{\partial r}\right)-\frac{1}{r^2}\frac{\partial^2 (r\gamma_{r\theta}^0)}{\partial r\partial\theta} = \frac{\nabla^2 w}{R} \tag{269}$$

where ∇^2 is the Laplacian in polar coordinates. If we use the stress-strain law (206), and the stress function definitions (268)—just as we did earlier for the cylinder—we get the equation of compatibility:

$$\nabla^4 F-\frac{Eh}{R}\nabla^2 w = -(1-\nu)\nabla^2\Phi. \tag{270}$$

Another relation between F, w can be derived from the remaining three equations of equilibrium. First we eliminate the shear forces, Q_r, Q_θ to yield one equation,

$$\frac{\partial}{\partial r}\left[\frac{\partial (rM_r)}{\partial r}+\frac{\partial M_{r\theta}}{\partial \theta}-M_\theta\right]+\frac{1}{r}\frac{\partial}{\partial \theta}\left[\frac{\partial (rM_{r\theta})}{\partial r}+\frac{\partial M_\theta}{\partial \theta}+M_{r\theta}\right]-\frac{r}{R}(N_r+N_\theta)-rq_n = 0. \tag{271}$$

Now the moments can be given in terms of the deflection $w(r, \theta)$ as

$$M_r = -D\left[\frac{\partial^2 w}{\partial r^2}+\nu\left(\frac{1}{r}\frac{\partial w}{\partial r}+\frac{1}{r^2}\frac{\partial^2 w}{\partial \theta^2}\right)\right]$$

$$M_\theta = -D\left[\frac{1}{r}\frac{\partial w}{\partial r}+\frac{1}{r^2}\frac{\partial^2 w}{\partial \theta^2}+\nu\frac{\partial^2 w}{\partial r^2}\right]$$

$$M_{r\theta} = -(1-\nu)D\frac{\partial}{\partial r}\left(\frac{1}{r}\frac{\partial w}{\partial \theta}\right) \tag{272}$$

so that now both the shear forces can be given as

$$Q_r = -D\frac{\partial}{\partial r}(\nabla^2 w), \quad Q_\theta = -D\frac{1}{r}\frac{\partial}{\partial \theta}(\nabla^2 w) \tag{273}$$

and the equation of equilibrium (272) can be manipulated into the form

$$D\nabla^4 w+\frac{1}{R}\nabla^2 F = -q_n-\frac{2\Phi}{R}\,. \tag{274}$$

Thus equations (270), (274) form the coupled system from which w, F are to be determined. Note the similarity to equations (209), (212) derived from the cylinder equations of Donnell. Here, as there, if we let $R \to \infty$ we obtain the plane-stress equations of elasticity from equation (270), and the equation for the deformation of a circular plate from equation (274). Thus, while the cylindrical shell equations may be viewed as those of a curved rectangular plate, the spherical cap equations are those of a curved circular plate. And again we note the role played by the curvature of the shell in coupling the stretching and bending problems.

We shall not delve into solutions of equations (270), (274). Suffice it to say that solutions are made "easy" by the appearance of the ∇^2 operator in repeated form in these equations. As usual, the homogeneous solutions are needed to analyze the effects of edge loading, while the surface loads are included with particular integrals. For details, the standard texts should be consulted. However, it is worth noting that for shallow caps, the Reissner theory is quite accurate, and very useful.

We have not given a complete discussion of shells of revolution. For example, we have not discussed other approaches to the general shell of revolution equations, e.g., the Blumenthal approximation, and other asymptotic integration schemes. Also, we have not considered the bending of cones or toroids, for example, nor have we even solved a great many spherical shell problems. However, we have given the basic equations, and hopefully we have exposed the fundamental ideas.

APPENDIX VIA

The shell mating problem

WITH the aid of the Geckeler solution for an edge-loaded sphere, and the influence coefficients given in Appendix VB for a cylinder we can now consider the shell mating problem. It was demonstrated in Chapter IV that a pressurized cylinder and a pressurized hemisphere, of the same radius, would expand different amounts under the same pressure. Thus, if we wished to cap a pressurized cylinder with a hemispherical cap, a shear force and a moment—discontinuity stresses—would have to be applied at the juncture to insure a displacement and rotation fit. For the sign convention shown

the cylinder edge displacement and rotation are

$$w_c = \frac{(2-\nu)PR^2}{2Eh_c} + \frac{2c}{E}\left(\frac{R}{h_c}\right)^{3/2} Q_0 + \frac{2c^2}{E}\left(\frac{R}{h_c^2}\right) M_0$$

$$\beta_c = \frac{2c^2}{E}\left(\frac{R}{h_c^2}\right) Q_0 + \frac{4c^3}{E}\left(\frac{R^{1/2}}{h_c^{5/2}}\right) M_0$$

and for the sphere

$$w_s = \frac{(1-\nu)PR^2}{2Eh_s} - \frac{2c}{E}\left(\frac{R}{h_s}\right)^{3/2} Q_0 + \frac{2c^2}{E}\left(\frac{R}{h_s^2}\right) M_0$$

$$\beta_s = \frac{2c^2}{E}\left(\frac{R}{h_s^2}\right) Q_0 - \frac{4c^3}{E}\left(\frac{R^{1/2}}{h_s^{5/2}}\right) M_0$$

where $c^4 = 3(1-\nu^2)$ and h_c and h_s are the cylinder and sphere wall thicknesses, respectively. We assume the same radius, modulus and Poisson's ratio for both shells.

For a smooth, continuous joining of the shells, we require that at the juncture

$$w_s = w_c, \quad \beta_s = \beta_c$$

so that Q_0 and M_0 are determined by the equations

$$\frac{2c^2}{E}\left(\frac{R}{h_s}\right)^2\left[1-\left(\frac{h_s}{h_c}\right)^2\right]\frac{M_0}{R}-\frac{2c}{E}\left(\frac{R}{h_s}\right)^{3/2}\left[1+\left(\frac{h_s}{h_c}\right)^{3/2}\right]Q_0$$
$$=\frac{Ph_s}{2E}\left(\frac{R}{h_s}\right)^2\left[(2-\nu)\left(\frac{h_s}{h_c}\right)-(1-\nu)\right]$$

and

$$\frac{2c^2}{E}\left(\frac{R}{h_s}\right)^2\left[\left(\frac{h_s}{h_c}\right)^2-1\right]Q_0+\frac{4c^3}{E}\left(\frac{R}{h_s}\right)^{5/2}\left[1+\left(\frac{h_s}{h_c}\right)^{5/2}\right]\frac{M_0}{R}=0.$$

Notice, from the second equation, which enforces slope compatibility, that if the shell thicknesses are equal ($h_s = h_c$) then the *discontinuity moment* M_0 vanishes! For that case, compatibility of both slope and displacement is enforced simply through the *discontinuity shear force* Q_0 acting alone.

Consider a simple problem, taken from an aerospace calculation (check the membrane stresses due to the pressure!). Let

$$R = 50 \text{ in.}, \quad h_s = 0.05 \text{ in.}, \quad h_c = 0.10 \text{ in.}$$

$$E = 28\times 10^6 \text{ psi}, \quad \nu = 0.30, \quad P = 300 \text{ psi}.$$

Then a straightforward calculation yields

$$Q_0 = 70.4 \text{ pounds/inch}$$
$$M_0 = 27.6 \text{ pound-inches/inch}.$$

CHAPTER VII

Axisymmetric Vibrations of Circular Cylinders

VIII-1. FREE VIBRATIONS—FREQUENCIES, MODE SHAPES, ORTHOGONALITY

This chapter will be devoted to a discussion—largely by example—of the *free and forced vibrations of shell structures.* We will indicate the complete free vibration problem, approximations for "radial" frequencies, orthogonality of free vibration modes, and two types of normal mode expansions for the forced vibrations of shells.

Consider first the axisymmetric vibrations of circular cylinders, using the Donnell equations to describe these vibrations, i.e.,

$$\frac{\partial^2 u}{\partial x^2}+\frac{\nu}{R}\frac{\partial w}{\partial x} = \frac{1-\nu^2}{E}\,\rho\,\frac{\partial^2 u}{\partial t^2}-\frac{1-\nu^2}{Eh}\,q_x(x, t)$$

$$\frac{\nu}{R}\frac{\partial u}{\partial x}+\frac{h^2}{12}\frac{\partial^4 w}{\partial x^4}+\frac{w}{R^2} = -\frac{1-\nu^2}{E}\,\rho\,\frac{\partial^2 w}{\partial t^2}+\frac{1-\nu^2}{Eh}\,q(x, t) \qquad (275)$$

where $q(x, t) = -q_n(x, t)$. The equations of motion (275) can be used to look at the response of cylinders to the (time-dependent) axial shear load $q_x(x, t)$ and radial pressure load $q(x, t)$. Let us examine first the *free vibrations* of a cylindrical shell, subject to the simple support boundary conditions

$$w = \frac{\partial^2 w}{\partial x^2} = N_x = 0 \quad \text{at} \quad x = 0, L. \qquad (276)$$

In this instance, after setting $q_x = q = 0$, we seek solutions to the homogeneous partial differential equations (275) subject to the homogeneous boundary conditions (276). Thus we are seeking a set of *eigenfunctions* corresponding to a set of *eigenvalues* that allow a non-trivial solution to the homogeneous system described above. For this system, for periodic oscillations, we can write the solutions as[22]

$$u(x, t) = U(x) \cos \omega t, \quad w(x, t) = W(x) \cos \omega t \tag{277}$$

where

$$U(x) = U_m \cos \frac{m\pi x}{L}, \quad W(x) = W_m \sin \frac{m\pi x}{L} \tag{278}$$

where $m = 0, 1, 2 \ldots$ These solutions will satisfy the boundary conditions (276), and the differential equations if

$$-\left(\frac{m\pi}{L}\right)^2 U_m + \frac{\nu}{R}\left(\frac{m\pi}{L}\right) W_m = -\frac{1-\nu^2}{E} \rho\omega^2 U_m$$

$$-\frac{\nu}{R}\left(\frac{m\pi}{L}\right) U_m + \left[\frac{1}{R^2} + \frac{h^2}{12}\left(\frac{m\pi}{L}\right)^4\right] W_m = \frac{1-\nu^2}{E} \rho\omega^2 W_m$$

or

$$\begin{bmatrix} (K^2-\lambda^2) & \nu\lambda \\ \nu\lambda & (K^2-1-H\lambda^4) \end{bmatrix} \begin{Bmatrix} U_m \\ W_m \end{Bmatrix} = 0. \tag{279}$$

In equation (279) we have introduced

$$\lambda = \frac{m\pi R}{L} = \text{axial wavelength parameter}$$

$$H = \frac{1}{12}\left(\frac{h}{R}\right)^2 = \text{thickness parameter } (= k^2) \tag{280}$$

$$K^2 = \frac{(1-\nu^2)\rho}{E} R^2\omega^2 = (\text{frequency factor})^2.$$

[22] One can demonstrate this by allowing $U(x) = e^{px}$, etc., or by referring back to the analogy between beam bending and axisymmetric shell deformation and extrapolating beam eigenfunctions.

The homogeneous algebraic system (279) has a non-trivial solution only if the determinant of the coefficients vanishes, i.e., only if numbers K can be found that satisfy

$$K^4-(1+H\lambda^4+\lambda^2)K^2+\lambda^2(1-\nu^2+H\lambda^4) = 0. \tag{281}$$

The quadratic equation (281) is easily solved—some results are given on p. 135 in tabular form—to obtain the two frequency factors K_1, K_2 that correspond (for a given shell geometry) to a given value of the axial wave number m. With these results, the ratio of the coefficients of the axial and radial displacements may be calculated, i.e.,

$$\frac{U_m^{(r)}}{W_m^{(r)}} = -\frac{\nu\lambda}{K_r^2-\lambda^2}, \quad r = 1, 2. \tag{282}$$

Some results for these ratios are also given in the tables, and it is clear that there are clearly defined variations in this ratio with the frequency factor K_r. Thus, for $\lambda \leqq 1$, we see that $K_1 \cong 1$, and that the ratio (282) for $r = 1$ is always less than unity. For $\lambda > 1$ the pattern is reversed, and the ratio of displacement coefficients is less than unity for $r = 2$, and here $K_2 \cong 1$. What this signifies—in part—is that we can separate out motion that is predominantly radial in character by looking for those regions where the ratio (282) is significantly smaller than unity. Similarly, for those regions where the ratio is larger than unity, the motion is largely in the plane of the shell middle surface, in the axial direction. This suggests that for those instances where the shell loading is radial, and where the shell motion would be predominantly radial, we might seek an approximation to the above results by using this information to simplify the problem.

Before proceeding with such an approximation, it is useful to do some elementary energy calculations. Let us examine the strain energy during the free oscillations. For axisymmetric plane stress, the strain energy is

$$S = \frac{E}{2(1-\nu^2)}\int_0^L\int_0^{2\pi R}\int_{-h/2}^{h/2}(\varepsilon_{xx}^2+2\nu\varepsilon_{xx}\varepsilon_{yy}+\varepsilon_{yy}^2)dzdydx$$

TABLE OF APPROXIMATE FREQUENCY FACTOR (K), EXACT FREQUENCY FACTORS (K_1, K_2), DISPLACEMENT RATIOS AND STRAIN ENERGY FOR $h/R = 0.050$

AXISYMMETRIC FREE VIBRATIONS

λ	K	K_1	K_2	$(U/W)_1$	$(U/W)_2$	E_{1m}	E_{2m}	E_b
0.125	0.953939	1.000712	0.119157	−0.038040	26.287781	1.002874	9.826033	0.000000
0.250	0.953940	1.002976	0.237777	−0.079494	12.579576	1.012319	9.003422	0.000001
0.375	0.953941	1.007217	0.355165	−0.128739	7.767705	1.031296	7.737190	0.000004
0.500	0.953946	1.014346	0.470227	−0.192580	5.192725	1.067045	6.183280	0.000013
0.625	0.953956	1.026206	0.580997	−0.283030	3.533222	1.137426	4.551471	0.000032
0.750	0.953974	1.046477	0.683703	−0.422444	2.367179	1.290483	3.086759	0.000066
0.875	0.954003	1.081836	0.771607	−0.648559	1.541876	1.662537	2.010696	0.000122
1.000	0.954048	1.140221	0.836722	−0.999656	1.000347	2.599105	1.400486	0.000208
1.125	0.954114	1.222950	0.877696	−1.467503	0.581439	4.716166	1.127732	0.000334
1.250	0.954206	1.322913	0.901614	−1.998932	0.500271	8.742523	1.015845	0.000509
1.375	0.954329	1.432627	0.915942	−2.549541	0.392231	15.392737	0.967273	0.000745
1.500	0.954492	1.547657	0.925100	−3.098241	0.322768	25.386368	0.943912	0.001055
1.625	0.954700	1.665729	0.931357	−3.637339	0.274927	39.482483	0.931537	0.001453
1.750	0.954963	1.785651	0.935897	−4.165013	0.240099	58.499466	0.924442	0.001954
1.875	0.955288	1.906773	0.939370	−4.681364	0.213618	83.312012	0.920107	0.002575
2.000	0.955685	2.028711	0.942160	−5.187245	0.192781	114.854691	0.917321	0.003333
2.125	0.956163	2.151228	0.944505	−5.683953	0.175933	154.134720	0.915455	0.004248
2.250	0.956734	2.274171	0.946565	−6.172665	0.162006	202.223312	0.914162	0.005339
2.375	0.957407	2.397436	0.948448	−6.654229	0.150282	260.241943	0.913240	0.006628
2.500	0.958195	2.520951	0.950232	−7.129482	0.140264	329.378418	0.912567	0.008138

which for the axisymmetric Donnell theory becomes

$$S = \pi RC \int_0^L \left[\left(\frac{\partial u}{\partial x}\right)^2 + 2\nu \frac{w}{R}\frac{\partial u}{\partial x} + \left(\frac{w}{R}\right)^2 + \frac{h^2}{12}\left(\frac{\partial^2 w}{\partial x^2}\right)^2\right] dx. \tag{283}$$

Using equations (277), (278), the orthogonality of sines and cosines, and letting $\cos^2\omega t$ be replaced by unity, we can find that the *membrane strain energy* is given by[23]

$$S_{\text{membrane}} = \frac{\pi EL}{2(1-\nu^2)}\left(\frac{h}{R}\right)\left(W_m^{(r)}\right)^2 \left\{1 - 2\nu\lambda \frac{U_m^{(r)}}{W_m^{(r)}} + \left(\frac{U_m^{(r)}}{W_m^{(r)}}\right)^2 \lambda^2\right\} \tag{284}$$

while the *bending strain energy* is

$$S_{\text{membrane}} = \frac{\pi EL}{2(1-\nu^2)}\left(\frac{h}{R}\right)\left(W_m^{(r)}\right)^2 \{H\lambda^4\}. \tag{285}$$

The dimensionless parts of these energy expressions that appear in the braces above are also given in the tables of results for this problem. We see first of all that the bending strain energy is considerably smaller —in fact almost negligibly so—than the energy due to stretching. In addition, note that there is a direct correspondence between the relative size of the stretching energy (for $r = 1$ against $r = 2$) and the displacement ratio (282). This, of course, is not very surprising, for the axial displacement occurs only in the membrane energy term.

Now if we desire an approximation for motion that is predominantly radial, we obtain it by *deleting the in-plane inertia*. In equations (279) this is equivalent to setting $K^2 = 0$ in the (1, 1) term in the matrix of coefficients, leading immediately to the (approximate) *radial frequency*

$$K^2 = 1 - \nu^2 + H\lambda^4. \tag{286}$$

This approximate frequency is also tabulated, and it can be seen to be a reasonable approximation to the frequency that corresponds to motion that is predominantly radial.

[23] Calculations of this type—although considerably more extensive—were apparently originally given in the classic papers of Arnold and Warburton. See the Reference list at the end of the chapter (p. 150).

It is also of interest to note that for $\lambda > 1$, the lowest frequency of the exact pair is that corresponding to radial motion, while for $\lambda < 1$ the lowest frequency corresponds to motion that is largely axial. Investigations of non-symmetric vibrations have not found this phenomenon. In general, for a pair of axial and circumferential wave numbers, m and n, the frequency of the motion that is predominantly radial is an order of magnitude lower than that corresponding to axial motion, which is generally smaller than that corresponding to torsional (or circumferential) motion.[24] Since one is generally interested in the lowest frequencies that might be excited by a forcing function of one type or another, we have added impetus to seek approximations to the radial frequency.

It is left as an exercise to show that for the asymmetric vibrations of circular cylinders, use of the modal solutions for simple supports,

$$\begin{aligned} u &= A \cos \frac{m\pi x}{L} \cos \frac{ny}{R} \cos \omega t \\ v &= B \sin \frac{m\pi x}{L} \sin \frac{ny}{R} \cos \omega t \\ w &= C \sin \frac{m\pi x}{L} \cos \frac{ny}{R} \cos \omega t \end{aligned} \tag{287}$$

together with the Donnell shell theory [see equations (156)], yields the frequency determinant

$$\begin{bmatrix} \left(K^2-\lambda^2-\frac{1-\nu}{2}n^2\right) & \frac{1+\nu}{2}n\lambda & -\nu\lambda \\ \frac{1+\nu}{2}n\lambda & \left(K^2-\frac{1-\nu}{2}\lambda^2-n^2\right) & n \\ -\nu\lambda & n & [K^2-1-H(\lambda^2+n^2)^2] \end{bmatrix} \times$$

$$\times \begin{Bmatrix} A \\ B \\ C \end{Bmatrix} = 0. \tag{288}$$

[24] It ought not to be difficult to see that by extension of equations (275), (279) to non-symmetric oscillations, a triplet of frequencies occurs for each pair of values (m, n). See, for example, Refs. 1, 2, 3, 5, 6 in the list at the end of the chapter (p. 150).

An approximate solution for the radial frequency can be obtained by deleting the K^2 term in the first two of equations (288), or, alternatively, by deleting the in-plane inertia terms in the equations of motion, and then uncoupling these equations as we did in Chapter V. This would yield the equation of motion

$$\frac{h^2}{12}\nabla^8 w + \frac{1-\nu^2}{R^2}\frac{\partial^4 w}{\partial x^4} + \frac{\rho(1-\nu^2)}{E}\nabla^4\frac{\partial^2 w}{\partial t^2} = 0. \tag{289}$$

Using the last of the modal solutions (287) yields the radial frequency factor

$$K^2 = \frac{(1-\nu^2)\lambda^4 + H(\lambda^2+n^2)^4}{(\lambda^2+n^2)^2}. \tag{290}$$

For $n = 0$, the axisymmetric case, equations (290), (286) are in complete agreement, as they should be. It is now a simple numerical task to show that for a variety of values of λ, n, that there are three roots to equation (288) for each (λ, n) pair, that the lowest of each triplet is well approximated by equation (290), and that for the lowest of the triplet, ratios A/C and B/C are rather small. Some tables of these results are the given on pp. 140–1.

It is of interest to note, in these results, that for a given value of λ, the two higher frequencies increase monotonically with the circumferential wave number, while the low "radial" frequency does not. This behavior was explained by Arnold and Warburton in the extended version of the energy analysis outlined earlier. They showed that the stretching (membrane) strain energy decreased sharply with increasing circumferential wave number, followed by an increase in the bending strain energy. A schematic is given in Fig. 14.

That schematic quickly reveals that as the number of circumferential nodes increases, there is a transition from behavior that is largely stretching in the shell middle surface to behavior that reflects bending of the shell wall. It is also seen that the stretching energy is sensitive to the number of axial waves as well as the number of circumferential waves, while the bending energy is sensitive only to the number of circumferential nodes.

Thus, to sum up what we have seen so far in our study of the free vibrations of cylinders, each pair of wave numbers produces three frequencies which correspond in turn to motion of the shell middle surface that is predominantly radial, axial and circumferential, respectively, for the increasing frequencies in each triplet. Also, a very satisfactory approximation to the low frequency can be obtained by deleting the in-plane inertia terms in equations of motion.

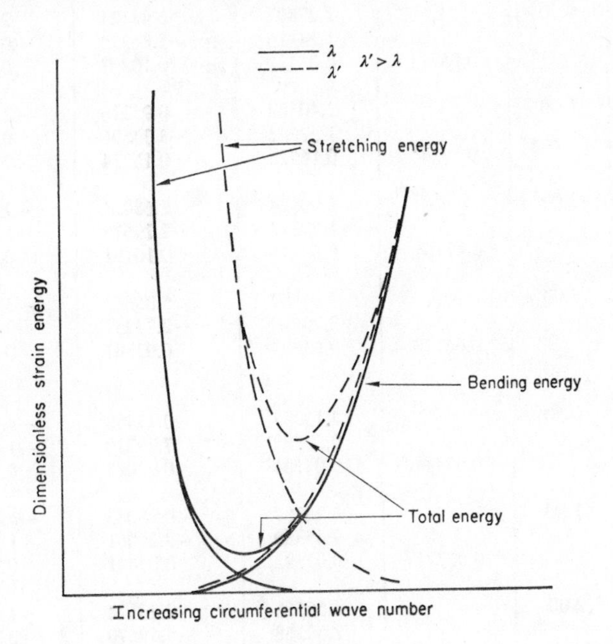

FIG. 14.

Before leaving the study of the free vibrations of cylinders, a few remarks on the subject of *orthogonality* are mandatory. In particular we shall indicate that for each frequency there is a discrete mode shape that is orthogonal in a particular sense to every other mode shape corresponding to every other frequency of free vibration. Not only are the eigenfunctions corresponding to different wave numbers orthogonal, but for the three (the general case) frequencies that correspond to a given pair of wave numbers (λ, n), each has a mode shape that is orthogonal to the other two. To illustrate this, consider the pair of

TABLE OF RADIAL FREQUENCY FACTORS, EXACT FREQUENCIES, AND CORRESPONDING DISPLACEMENT RATIOS

n	λ	K	K_i	$(A/C)_i$	$(B/C)_i$
2	0.50		2.27807	0.42124	−2.06313
			1.24316	−5.86278	−0.71233
		0.05771	0.05123	0.10897	0.50695
2	1.00		2.40589	0.95226	−2.25120
			1.39042	−3.33596	−0.96691
		0.19144	0.16928	0.15234	0.50865
2	2.00		2.90924	2.68024	−2.92743
			1.77342	−2.29673	−1.76120
		0.47764	0.43816	0.10191	0.43490
2	3.00		3.64482	5.35633	−3.73117
			2.20598	−2.23257	−2.93699
		0.66170	0.63293	0.03301	0.31540
4	0.50		4.15097	0.47458	−4.03918
			2.38892	−17.79710	−1.84346
		0.05344	0.05181	0.03018	0.25112
4	1.00		4.23387	0.98083	−4.15713
			2.45389	−9.29061	−1.19515
		0.07771	0.07523	0.05441	0.25339
4	2.00		4.55558	2.20174	−4.60707
			2.68568	−5.39570	−2.36158
		0.20100	0.19438	0.07471	0.25276
4	3.00		5.05700	3.83443	−5.27901
			3.90950	−3.39673	−3.00415
		0.35240	0.34231	0.06550	0.23701
6	0.50		6.10255	0.48790	−6.02580
			3.56324	−37.75770	−2.89120
		0.11482	0.11342	0.01369	0.16706
6	1.00		6.16152	0.99017	−6.10893
			3.60353	−19.26190	−2.95839
		0.11981	0.11812	0.02612	0.16793

TABLE (*cont.*)

n	λ	K	K_i	$(A/C)_i$	$(B/C)_i$
6	2.00		6.39293	2.09224	−6.43303
			3.75794	−10.38830	−3.22319
		0.15843	0.15606	0.04363	0.16964
6	3.00		6.76359	3.40200	−6.94401
			3.99615	−7.75334	−3.65449
		0.23801	0.23444	0.04964	0.16833
8	0.50		8.07746	0.49283	−8.01626
			4.74266	−65.67430	−3.91283
		0.20321	0.20163	0.00777	0.12522
8	1.00		8.12279	0.99379	−8.07991
			4.77187	−33.21590	−3.96166
		0.20607	0.20444	0.01512	0.12562
8	2.00		8.30190	2.05161	−8.33055
			4.88632	−17.36250	−4.15592
		0.22223	0.22039	0.02726	0.12675
8	3.00		8.59292	3.23196	−8.73452
			5.06943	−12.40760	−4.47658
		0.25908	0.25686	0.03466	0.12731
10	0.50		10.06220	0.49507	−10.00730
			5.92376	−101.50400	−4.92160
		0.31703	0.31545	0.00499	0.10017
10	1.00		10.09890	0.99536	−10.05870
			5.94673	−51.12810	−4.96000
		0.31953	0.31792	0.00982	0.10038
10	2.00		10.24440	2.03206	−10.26210
			6.03748	−26.31520	−5.11321
		0.33092	0.32919	0.01836	0.10108
10	3.00		10.48280	3.14885	−10.59370
			6.18507	−18.37520	−5.37640
		0.35357	0.35165	0.02470	0.10174

equations corresponding to the pth eigenvalue ω_p, with corresponding mode shapes U_p, W_p, so that from equations (275), (277) we can write

$$U_p''+\frac{\nu}{R}W_p'-\frac{1-\nu^2}{E}\rho\omega_p^2U_p=0$$
$$\frac{\nu}{R}U_p'+\frac{h^2}{12}W_p^{\mathrm{IV}}+\frac{1}{R^2}W_p+\frac{1-\nu^2}{E}\rho\omega_p^2W_p=0. \tag{291}$$

Now equations (291) must also be true for the frequency ω_q with its corresponding mode shapes, so that now we can form the following integral expression[25]

$$\int_0^L\Big\{\Big[U_p''+\frac{\nu}{R}W_p'-\frac{1-\nu^2}{E}\rho\omega_p^2U_p\Big]U_q$$
$$+\Big[-\frac{\nu}{R}U_p'-\frac{h^2}{12}W_p^{\mathrm{IV}}-\frac{1}{R^2}W_p-\frac{1-\nu^2}{E}\rho\omega_p^2W_p\Big]W_q$$
$$-\Big[U_q''+\frac{\nu}{R}W_q'-\frac{1-\nu^2}{E}\rho\omega_q^2U_q\Big]U_p \tag{292}$$
$$-\Big[-\frac{\nu}{R}U_q'-\frac{h^2}{12}W_q^{\mathrm{IV}}-\frac{1}{R^2}W_q-\frac{1-\nu^2}{E}\rho\omega_q^2W_q\Big]W_p\Big\}dx=0.$$

Since the eigenfunctions satisfy equations (291), the integral (292) must be zero. If the terms with derivatives are integrated by parts, e.g.,

$$\int_0^L[U_p''U_q-U_q''U_p]dx=[U_p'U_q-U_q'U_p]_{x=0}^{x=L}$$

and the boundary conditions satisfied for all the eigenfunctions, one can then find that

$$(\omega_p^2-\omega_q^2)\int_0^L(U_pU_q+W_pW_q)dx=0. \tag{293}$$

[25] The reader may recall here a process similar to that of proving the Maxwell–Betti reciprocal theorem for shells. Using those results makes the above calculation much simpler.

Equation (293) is the orthogonality condition for the axisymmetric vibrations of circular cylinders. It states that for two modes such that $\omega_p \neq \omega_q$, the mode shapes must satisfy

$$\int_0^L (U_p U_q + W_p W_q) dx = 0. \tag{294}$$

For the problem considered earlier—the simple support boundary conditions—we would have

$$U_p(x) = U_p^{(r)} \cos \frac{p\pi x}{L}, \quad W_p(x) = W_p^{(r)} \sin \frac{p\pi x}{L}$$

so that the orthogonality condition would read

$$\int_0^L \left(U_p^{(r)} U_q^{(s)} \cos \frac{p\pi x}{L} \cos \frac{q\pi x}{L} + W_p^{(r)} W_q^{(s)} \sin \frac{p\pi x}{L} \sin \frac{q\pi x}{L} \right) dx = 0. \tag{295}$$

For different wave numbers p, q, the above identity is always satisfied because of the orthogonality of the trigonometric functions. Thus for $p \neq q$ the orthogonality condition is "trivially" satisfied.

However, if $p = q$, the only way that we can have $\omega_p \neq \omega_q$ is if $r \neq s$. In this instance the orthogonality relation (295) reduces to

$$U_p^{(r)} U_p^{(s)} + W_p^{(r)} W_p^{(s)} = 0, \quad r \neq s$$

or

$$\frac{U_p^{(r)}}{W_p^{(r)}} \frac{U_p^{(s)}}{W_p^{(s)}} = -1. \tag{296}$$

For this problem the orthogonality condition (296) can be (analytically) shown to be satisfied by noting that equation (281) can be written as $(K^2 - K_r^2)(K^2 - K_s^2)$ so that we can identify

$$K_r^2 + K_s^2 = 1 + \lambda^2 + H\lambda^4, \quad K_r^2 K_s^2 = \lambda^2(1 - \nu^2 + H\lambda^4).$$

Using now the ratio (282) we may formulate the left-hand side of the condition (296) as

$$\frac{\nu^2\lambda^2}{(K_r^2-\lambda^2)(K_s^2-\lambda^2)} = \frac{\nu^2\lambda^2}{\lambda^2(1-\nu^2+H\lambda^4)-\lambda^2(1+H\lambda^4+\lambda^2)+\lambda^4}$$

$$= \frac{\nu^2\lambda^2}{-\nu^2\lambda^2}$$

so that orthogonality is indeed satisfied. As a further corroboration we have given in the tables earlier the ratios (282) for $r = 1, 2$. It is a simple matter to find that their respective products are always equal to minus one.

As a final point on the mode shapes, we note that as they are the "solutions" of an eigenvalue problem, they all contain an arbitrary coefficient. While this is of no consequence in a free vibration problem, it is useful to normalize the mode shapes for forced vibration analyses. Thus we will write the orthogonality condition (294) as an *orthonormality* condition

$$\int_0^L (U_pU_q+W_pW_q)dx = \delta_{pq}N_p \tag{297}$$

where δ_{pq} is the Kronecker delta, and it is equal to unity if p, q represent the same frequency and mode shape, and it vanishes for different frequencies.

VII-2. FORCED VIBRATIONS—NORMAL MODES SOLUTION

We now turn to the problem of obtaining the dynamic response of a shell to an applied dynamic loading, that is, we are now going to consider the *forced vibration* problem. There are two types of problems that we shall consider—the problem of applied surface loading with homogeneous boundary conditions, and then the problem of transient edge loading on a shell. In both instances we shall seek a *normal modes solution*, that is, a response whose spatial variation may be represented in terms of the normal modes of free vibration. Thus we write, for the axisymmetric response of the simply supported circular cylinder,

$$
\begin{aligned}
u(x, t) &= \sum_{m=1}^{\infty} \sum_{r=1}^{2} U_m^{(r)} \cos \frac{m\pi x}{L} q_m^{(r)}(t) \\
w(x, t) &= \sum_{m=1}^{\infty} \sum_{r=1}^{2} W_m^{(r)} \sin \frac{m\pi x}{L} q_m^{(r)}(t).
\end{aligned}
\tag{298}
$$

Substitution of the expansions (298) into the partial differential equations (275) yields:

$$
\sum_{m=1}^{\infty} \sum_{r=1}^{2} \left\{\left[\frac{1-\nu^2}{E} \rho \ddot{q}_m^{(r)} + \left(\frac{m\pi}{L}\right)^2 q_m^{(r)}\right] U_m^{(r)} - \frac{\nu}{R} \frac{m\pi}{L} q_m^{(r)} W_m^{(r)}\right\} \cos \frac{m\pi x}{L} = \frac{1-\nu^2}{Eh} q_x(x, t) \tag{299}
$$

and

$$
\sum_{m=1}^{\infty} \sum_{r=1}^{2} \left\{\left[\frac{1-\nu^2}{E} \rho \ddot{q}_m^{(r)} + \left(\frac{h^2}{12}\left(\frac{m\pi}{L}\right)^4 + \frac{1}{R^2}\right) q_n^{(r)}\right] W_m^{(r)} - \frac{\nu}{R}\left(\frac{m\pi}{L}\right) q_m^{(r)} U_m^{(r)}\right\} \sin \frac{m\pi x}{L} = \frac{1-\nu^2}{Eh} q(x, t). \tag{300}
$$

Now the displacement coefficients $U_m^{(r)}$, $W_m^{(r)}$ must satisfy the algebraic equations (279)—or their immediate antecedents—so that equations (299), (300) can be written as

$$
\sum_{m=1}^{\infty} \sum_{r=1}^{2} \left[\ddot{q}_m^{(r)} + \omega_{mr}^2 q_m^{(r)}\right] U_m^{(r)} \cos \frac{m\pi x}{L} = \frac{1}{\rho h} q_x(x, t) \tag{301}
$$

$$
\sum_{m=1}^{\infty} \sum_{r=1}^{2} \left[\ddot{q}_m^{(r)} + \omega_{mr}^2 q_m^{(r)}\right] W_m^{(r)} \sin \frac{m\pi x}{L} = \frac{1}{\rho h} q(x, t). \tag{302}
$$

Now we multiply the first of the above by $U_n^{(s)} \cos (n\pi x/L)$, the second by $W_n^{(s)} \sin (n\pi x/L)$, and then add the two, and integrate over the shell length. In the process we then use our orthonormality relations to find that the *generalized coordinates* are governed by

$$
\ddot{q}_n^{(s)} + \omega_{ns}^2 q_n^{(s)} = \frac{1}{\rho h} Z_{ns}(t) \tag{303}
$$

where

$$Z_{ns}(t) = \frac{1}{N_n^{(s)}} \int_0^L \left(q_x U_n^{(s)} \cos \frac{n\pi x}{L} + q W_n^{(s)} \sin \frac{n\pi x}{L} \right) dx \tag{304}$$

and

$$N_n^{(s)} = \int_0^L \left[(U_n^{(s)})^2 \cos^2 \frac{n\pi x}{L} + (W_n^{(s)})^2 \sin^2 \frac{n\pi x}{L} \right] dx$$

$$= \frac{L}{2} [(U_n^{(s)})^2 + (W_n^{(s)})^2]. \tag{305}$$

The solution to equation (303) can be easily obtained through standard techniques, and it appears as

$$q_n^{(s)}(t) = q_n^{(s)}(0) \cos \omega_{ns} t + \frac{1}{\omega_{ns}} \dot{q}_n^{(s)}(0) \sin \omega_{ns} t$$

$$+ \frac{1}{\omega_{ns}} \int_0^t \frac{1}{\rho h} Z_{ns}(\tau) \sin [\omega_{ns}(t-\tau)] d\tau. \tag{306}$$

The first two terms of this solution represent the *initial conditions*, given as

$$q_n^{(s)}(0) = \frac{1}{N_n^{(s)}} \int_0^L \left(u(x, o) U_n^{(s)} \cos \frac{n\pi x}{L} + w(x, o) W_n^{(s)} \sin \frac{n\pi x}{L} \right) dx \tag{307a}$$

$$\dot{q}_n^{(s)}(0) = \frac{1}{N_n^{(s)}} \int_0^L \left(\frac{\partial u(x, o)}{\partial t} U_n^{(s)} \cos \frac{n\pi x}{L} + \frac{\partial w(x, o)}{\partial t} W_n^{(s)} \sin \frac{n\pi x}{L} \right) dx. \tag{307b}$$

Thus, in principle, we have the complete modal solution for the response of a circular cylinder subject to axisymmetric surface loads. Before leaving this section, however, we shall use the particular portion of the solution (306), (304) to define *modal participation factors*. These factors give an indication of which modes are excited—or are likely to be excited—and which will be dominant in the response. For the present definition, we assume that the temporal portion of the excitation can be separated from the spatial, i.e.,

$$\begin{aligned} q_x(s, t) &= f_1(t)q_x^*(x) \\ q(x, t) &= f_2(t)q^*(x). \end{aligned} \tag{308}$$

With this assumption it is a straightforward manipulation to show that the particular response can be written in the form

$$q_n^{(s)}(t) = \left(\frac{1}{\rho h \omega_{ns}}\right)[m_{1ns}F_{1ns}(t) + m_{2ns}F_{2ns}(t)] \tag{309}$$

where we have introduced the time-dependent results

$$F_{ins}(t) = \int_0^t f_i(\tau) \sin[\omega_{ns}(t-\tau)]d\tau, \quad i = 1, 2 \tag{310}$$

and the *modal participation factors*[26]

$$\begin{aligned} m_{1ns} &= \frac{1}{N_n^{(s)}} \int_0^L q_x^*(x) U_n^{(s)} \cos\frac{n\pi x}{L}\, dx \\ m_{2ns} &= \frac{1}{N_n^{(s)}} \int_0^L q^*(x) W_n^{(s)} \sin\frac{n\pi x}{L}\, dx. \end{aligned} \tag{311}$$

The participation factors (311) now serve to indicate whether the modal response will be largely axial or largely radial, which in turn is dictated by the relative size of $q_x^*(x)$ and $q^*(x)$, and of $U_n^{(s)}$ and $W_n^{(s)}$.

VII-3. FORCED VIBRATIONS—WILLIAMS' METHOD FOR TIME-DEPENDENT BOUNDARY CONDITIONS

As a final exercise, we shall now outline a modal solution for a problem with time-dependent boundary conditions, i.e., shells with applied edge loading that is time dependent. This response analysis will be trickier than the above, for the modal expansions satisfy only homogeneous conditions at the edges. However, making use of the principle

[26] Note that the definitions of m.p.f.s vary somewhat in the technical literature, but the idea behind all the definitions is as described above.

of superposition, we can develop a modal solution which is known as the *Williams modal acceleration technique.*

To illustrate the Williams solution we shall examine the radial response of a circular cylinder subject to the boundary conditions

$$w(\pm L/2, t) = 0, \quad M_x(\pm L/2, t) = -D\,\frac{\partial^2 w(\pm L/2, t)}{\partial x^2} = M_0 f(t). \tag{312}$$

To predict the radial response we shall ignore the in-plane inertia, and use the uncoupled, axisymmetric, Donnell equation,

$$\frac{h^2}{12}\frac{\partial^4 w}{\partial x^4}+\frac{1-\nu^2}{R^2}\,w = -\frac{(1-\nu^2)\rho}{E}\frac{\partial^2 w}{\partial t^2}\,. \tag{313}$$

To obtain a solution to the system (312), (313) we will assume that

$$w(x, t) = w^{(s)}(x, t)+\sum_{n=1,3,5\ldots}^{\infty} W_n \cos\frac{n\pi x}{L}\, q_n(t) \tag{314}$$

where $\cos(n\pi x/L)$ are the eigenfunctions of a simply supported shell, with the origin of coordinates at the center of the shell, and with the free vibration frequencies given by

$$\omega_n^2 = \frac{E}{(1-\nu^2)\rho R^2}\,(1-\nu^2+H\lambda^4) \tag{315}$$

The *static solution* $w^{(s)}(x, t)$ satisfies

$$H\frac{\partial^4 w^{(s)}}{\partial x^4}+(1-\nu^2)w^{(s)} = 0 \tag{316}$$

subject to the boundary conditions (312). Thus, to the modal solution that satisfies homogeneous boundary conditions we are adding a solution that is based on the static shell equation, but that satisfies the time-dependent boundary conditions. The inertia of this static solution will then be treated as a radial surface load that excites the normal modes. It is an exercise from beam theory to show that the static solution is given by, for this problem,

$$w^{(s)}(x, t) = \frac{M_0 R^2 f(t)}{4\phi^2 D}\left[\sinh\frac{\phi(L+2x)}{2R}\sin\frac{\phi(L-2x)}{2R}+\right.$$
$$\left.+\sinh\frac{\phi(L-2x)}{2R}\sin\frac{\phi(L+2x)}{2R}\right]\left[\sinh^2\frac{\phi L}{2R}+\cos^2\frac{\phi L}{2R}\right]^{-1} \tag{317}$$

where $4\phi^4 = (1-\nu^2)H$.

With $w^{(s)}(x, t)$ known, we now substitute the solution (314) into the differential equation (313), so that

$$\sum_{n=1,3,5...}^{\infty}\left[\frac{(1-\nu^2)\rho}{E}\ddot{q}_n(t)+\left(\frac{1-\nu^2}{R^2}+\frac{H}{R^2}\lambda^4\right)q_n(t)\right]W_n\cos\frac{n\pi x}{L}$$
$$= -\frac{(1-\nu^2)\rho}{E}\frac{\partial^2 w^{(s)}}{\partial t^2}. \tag{318}$$

or, in view of equation (315),

$$\sum_{n=1,3,5...}^{\infty}[\ddot{q}_n(t)+\omega_n^2 q_n(t)]W_n\cos\frac{n\pi x}{L} = -\frac{\partial^2 w^{(s)}}{\partial t^2}. \tag{319}$$

By applying the appropriate orthonormality condition we see that

$$\ddot{q}_n(t)+\omega_n^2 q_n(t) = \frac{1}{N_n}\ddot{Q}_n(t) \tag{320}$$

where

$$N_n = \int_{-L/2}^{L/2} W_n^2\cos^2\frac{n\pi x}{L}\,dx \tag{321}$$

and

$$Q_n(t) = -\int_{-L/2}^{L/2} W_n w^{(s)}(x, t)\cos\frac{n\pi x}{L}\,dx. \tag{322}$$

Finally, the solution to equation (320), for a shell initially at rest in the equilibrium position, is

$$q_n(t) = \frac{1}{N_n}[Q_n(t)-\omega_n\int_0^t Q_n(\tau)\sin\omega_n(t-\tau)d\tau]. \tag{323}$$

To complete the solution, it is a straightforward integration to evaluate the integral (322) by substitution from equation (317):

$$Q_n(t) = -\pi \left(\frac{N_n}{2L}\right)^{1/2} \frac{ER^2 n(-1)^{(n-1)/2}}{\rho D\phi^4 L\omega_n^2} M_0 f(t). \tag{324}$$

Thus, our solution is complete, in terms of equations (314), (317), (323), (324). Note, incidentally, the appearance of the ω_n^2 in the denominator, which in the general case augurs very well for the rapidity of convergence of the sum of the normal modes.

This also completes our discussion of shell dynamics. Needless to say, it is no more than an introduction to the methodology, the terminology, and some of the types of results that may be expected. For example, we have covered only one simple geometry, we have said nothing at all about thick shells that require consideration of transverse shear deformation and rotatory inertia, and have said little about the determination of approximate frequencies and mode shapes. Nevertheless, a beginning has been made, and the reader can now more easily peruse the References given for this chapter.

REFERENCES

1. R. N. Arnold and G. B. Warburton, Flexural vibrations of the walls of thin cylindrical shells having freely supported ends, *Proceedings of the Royal Society*, Series A, Vol. 197, 1949, p. 238.
2. R. N. Arnold and G. B. Warburton, The flexural vibrations of thin cylinders, *Proceedings of the Institution of Mechanical Engineers*, Vol. 167, 1953, p. 62.
3. C. L. Dym, Vibrations of pressurized orthotropic cylindrical membranes, *AIAA Journal*, Vol. 8, No. 4, 1970, p. 693.
4. C. L. Dym, Vibrations of pressurized orthotropic shells, *AIAA Journal*, Vol. 9, No. 6, 1971, p. 1201.
5. C. L. Dym, Some new results for the vibrations of circular cylinders, *Journal of Sound and Vibration*, Vol. 29, No. 2, 1973, p. 189.
6. K. Forsberg, A review of the analytical methods used to determine the modal characteristics of cylindrical shells, *Lockheed Missiles and Space Company* Technical Report No. 6-75-65-25, May 1965.
7. A. Kalnins, Effect of bending on vibrations of spherical shells, *Journal of the Acoustical Society of America*, Vol. 36, No. 1, 1964, p. 75.
8. H. Kraus, *Thin Elastic Shells*, John Wiley and Sons, New York, 1967.
9. R. D. Mindlin and L. E. Goodman, Beam vibrations with time-dependent boundary conditions, *Journal of Applied Mechanics*, Vol. 17, No. 4, 1950, p. 377.

10. W. Nowacki, *Dynamics of Elastic Systems*, Chapman and Hall, Ltd., London, 1963.
11. E. Reissner, On transverse vibrations of thin shallow elastic shells, *Quarterly of Applied Mathematics*, Vol. 13, No. 2, 1955, p. 169.
12. J. Sheng, The response of a thin cylindrical shell to a transient surface loading, *AIAA Journal*, Vol. 3, 1965, p. 701.
13. J. P. Wilkinson, Transient response of thin elastic shells, *Journal of the Acoustical Society of America*, Vol. 39, No. 5, Pt. 1, 1966, p. 895.
14. D. Williams, Displacements of a linear elastic system under a given transient load, *The Aeronautical Quarterly*, Vol. 1, 1949, p. 123.

Bibliography

(See also References on shell dynamics at end of Chapter VII, p. 150)

1. J. R. COLBOURNE, Approximate roots of Flugge's characteristic equations for the closed cylindrical shell, *Journal of Applied Mechanics*, Vol. 36, 1969, p. 352.
2. L. H. DONNELL, *Stability of Thin-walled Tubes under Torsion*, NACA TR 479, 1933.
3. C. L. DYM, On the buckling of cylinders in axial compression, *Journal of Applied Mechanics*, Vol. 40, No. 2, 1973.
4. C. L. DYM and I. H. SHAMES, *Solid Mechanics: A Variational Approach*, McGraw–Hill, New York, 1973.
5. W. FLÜGGE, *Stresses in Shells*, Springer, Berlin, 1960.
6. A. L. GOLDENVEIZER, *Theory of Elastic Thin Shells*, Pergamon Press, 1961.
7. W. C. GRAUSTEIN, *Differential Geometry*, Macmillan, New York, 1935.
8. N. J. HOFF, The accuracy of Donnell's equation, *Journal of Applied Mechanics*, Vol. 22, 1955, p. 329.
9. S. H. IYER and S. H. SIMMONDS, The accuracy of Donnell's theory for very high harmonic loading on closed cylinders, *Journal of Applied Mechanics*, Vol. 39, 1972, p. 836.
10. J. KEMPNER, Remarks on Donnell's equations, *Journal of Applied Mechanics*, Vol. 22, 1955, p. 117.
11. W. T. KOITER, A consistent first approximation in the general theory of thin elastic shells, *Proceedings of the Symposium on Theory of Thin Elastic Shells*, Delft, Netherlands, 1951 (see Ref. 12).
12. W. T. KOITER (Ed.), *Proceedings of I.U.T.A.M. Symposium on the Theory of Thin Elastic Shells*, North-Holland, Amsterdam, 1960.
13. H. KRAUS, *Thin Elastic Shells*, Wiley, New York, 1967.
14. E. KREYSZIG, *Differential Geometry*, University of Toronto Press, Toronto, 1959.
15. H. L. LANGHAAR, *Energy Methods in Applied Mechanics*, Wiley, New York, 1962.
16. H. L. LANGHAAR, *Foundations of Practical Shell Analysis*, unpublished notes, University of Illinois, 1962.
17. A. E. H. LOVE, *The Mathematical Theory of Elasticity*, Cambridge University Press, 1934.
18. K. M. MUSHTARI and K. Z. GALIMOV, *Nonlinear Theory of Thin Elastic Shells*, Published for NSF and NASA by Israel Program for Scientific Translation, O.T.S., U.S. Dept. of Commerce, Washington, 1957.

19. D. Muster (Ed.), *Proceedings of a Symposium on the Theory of Shells*, University of Houston, 1967.
20. P. M. Naghdi, On the theory of thin elastic shells, *Quarterly of Applied Mathematics*, Vol. 14, 1957, p. 369.
21. P. M. Naghdi, A survey of recent progress in the theory of elastic shells, *Applied Mechanics Reviews*, Vol. 9, No. 9, Sept. 1956.
22. W. Nash, Recent advances in the buckling of thin shells, *Applied Mechanics Reviews*, Vol. 13, No. 3, Mar. 1960.
23. F. Niordson (Ed.), *IUTAM Symposium—Theory of Thin Shells*, Springer Verlag, 1971.
24. V. V. Novozhilov, *Foundations of the Nonlinear Theory of Elasticity*, Graylock Press, Rochester, N.Y., 1953.
25. V. V. Novozhilov, *The Theory of Thin Shells*, Noordhoff, Groningen, Netherlands, 1959.
26. E. Reissner, A new derivation of the equations for the deflections of elastic shells, *American Journal of Mathematics*, Vol. 63, 1941, p. 177.
27. E. Reissner, On axisymmetrical deformation of thin shells of revolution, *Proceedings of Symposia in Applied Mathematics*, Vol. 3, 1950, p. 27.
28. E. Reissner, Stress–strain relations in the theory of thin elastic shells, *Journal of Mathematics and Physics*, Vol. 31, 1952, p. 109.
29. J. L. Sanders, Jr., *An Improved First-approximation Theory for Thin Shells*, NASA Technical Report R-24, 1959.
30. J. G. Simmonds, A set of simple, accurate equations for circular cylindrical shells, *International Journal of Solids and Structures*, 1966, p. 525.
31. J. J. Stoker, *Differential Geometry*, Wiley-Interscience, New York, 1969.
32. D. Struik, *Differential Geometry*, Addison-Wesley, Cambridge, Mass., 1950.
33. S. Timoshenko and J. N. Goodier, *Theory of Elasticity*, McGraw-Hill, New York, 1951.
34. S. Timoshenko and S. Woinowsky-Krieger, *Theory of Plates and Shells*, McGraw-Hill, New York, 1959.
35. S. Timoshenko and J. M. Gere, *Theory of Elastic Stability*, McGraw-Hill, New York, 1959.
36. V. Z. Vlasov, *General Theory of Shells and its Applications in Engineering*, NASA TT F-99, 1964.
37. J. H. Williams, Discussion of Ref. 1, *Journal of Applied Mechanics*, Vol. 37, 1970, p. 327.

Index